Interdisciplinary Applied Mathematics

Volume 31

Editors
S.S. Antman   J.E. Marsden
L. Sirovich   S. Wiggins

Geophysics and Planetary Sciences

Mathematical Biology
L. Glass, J.D. Murray

Mechanics and Materials
R.V. Kohn

Systems and Control
S.S. Sastry, P.S. Krishnaprasad

T0235497

Problems in engineering, computational science, and the physical and biological sciences are using increasingly sophisticated mathematical techniques. Thus, the bridge between the mathematical sciences and other disciplines is heavily traveled. The correspondingly increased dialog between the disciplines has led to the establishment of the series: *Interdisciplinary Applied Mathematics.*

The purpose of this series is to meet the current and future needs for the interaction between various science and technology areas on the one hand and mathematics on the other. This is done, firstly, by encouraging the ways that mathematics may be applied in traditional areas, as well as point towards new and innovative areas of applications; and, secondly, by encouraging other scientific disciplines to engage in a dialog with mathematicians outlining their problems to both access new methods and suggest innovative developments within mathematics itself.

The series will consist of monographs and high-level texts from researchers working on the interplay between mathematics and other fields of science and technology.

Interdisciplinary Applied Mathematics

Volumes published are listed at the end of the book

R.M. Samelson   S. Wiggins

Lagrangian Transport in Geophysical Jets and Waves

The Dynamical Systems Approach

Springer

Roger M. Samelson
College of Oceanic and
    Atmospheric Sciences
Oregon State University
Corvallis, OR 97331-5503
USA
rsamelson@coas.oregonstate.edu

Stephen Wiggins
Department of Mathematics
University of Bristol
Clifton, Bristol BS8 1TW
UK
s.wiggins@bristol.ac.uk

Series Editors
S.S. Antman
Department of Mathematics and
    Institute for Physical Science and
    Technology
University of Maryland
College Park, MD 20742
USA
ssa@math.umd.edu

J.E. Marsden
Control and Dynamical Systems
Mail Code 107-81
California Institute of Technology
Pasadena, CA 91125
USA
marsden@cds.caltech.edu

L. Sirovich
Laboratory of Applied Mathematics
Department of Biomathematics
Mt. Sinai School of Medicine
Box 1012
NYC 10029
USA
chico@camelot.mssm.edu

Stephen Wiggins
Department of Mathematics
University of Bristol
Clifton, Bristol BS8 1TW
UK
s.wiggins@bristol.ac.uk

Mathematics Subject Classification (2000): 34C35, 37J, 70D10

ISBN   978-1-4419-2204-5   e-ISBN   978-0-387-46213-4

Printed on acid-free paper.

9 8 7 6 5 4 3 2 1

springer.com

In memory of

H. Samelson, 1916–2005

and

R.E. Latimer, 1917–2001
L.F. Latimer, 1918–2004

Preface

The purpose of this book is to provide an accessible introduction to a new set of methods for the analysis of Lagrangian motion in geophysical flows. These methods were originally developed in the abstract mathematical setting of dynamical systems theory, through a geometric approach to differential equations that ultimately owes much to the insights of Poincaré (1892). In the 1980s and 1990s, researchers in applied mathematics and fluid dynamics recognized the potential of this approach for the analysis of fluid motion. Despite these developments and the existence of a substantial body of work on geophysical fluid problems in the dynamical systems and geophysical literature, no introductory text has been available that presents these methods in the context of geophysical fluid flow.

The text is meant to be accessible to geophysical fluid scientists and students familiar with the mathematics of ordinary (mostly) and partial (sometimes) differential equations. It assumes little or no prior knowledge of dynamical systems theory. An effort is made to explain concepts from a physical point of view, and to avoid the theorem and proof constructions that appear in dynamical systems texts. We hope that this book will prove useful to graduate students, research scientists, and educators in any branch of geophysical fluid science in which the motion and transport of fluid, and of materials carried by the fluid, is of interest. We hope that it will also prove interesting and useful to applied mathematicians who seek an introduction to an intriguing and rapidly developing area of geophysical fluid dynamics.

The primary material is organized into seven chapters. The first introduces the geophysical context and the mathematical models of geophysical fluid flow that are explored in subsequent chapters. The second and third cover the simplest case of steady flow, develop basic mathematical concepts and definitions, and touch on some important topics from the classical theory of Hamiltonian systems. The fundamental elements and methods of Lagrangian transport analysis in time-dependent flows that are the main subject of the book are described in the fourth, fifth, and sixth chapters. The seventh chapter gives a brief survey of some of the rapidly evolving research in geophysical fluid dynamics that makes use of this new approach. Some ancillary material, including a glossary and an introduction to numerical methods, is given in the appendices. In the text, *italics* indicate definitions, or terms that appear in the glossary.

The concepts and methods described in this book have been discovered and developed by many different individuals. We are grateful to these many people for their contributions, both direct and indirect, to this text. A look at the bibliography will give a useful, though necessarily incomplete, outline of this group. We are especially grateful to Achi Dosanjh for her enthusiasm, support, and patience, to Andrew Bennett for his encouragement and thorough comments on a first draft of the manuscript, and to David Reinert for his assistance with the figures. We would like also to acknowledge the support and encouragement of Reza Malek-Madani, Wen Masters (now at the National Science Foundation), and Manny Fiadeiro of the Office of Naval Research, who played a central role in forging links with diverse groups in oceanography and applied mathematics that led to the development of the essentially new approach to Lagrangian transport in geophysical flows described in this book.

RMS   SW
Corvallis, Oregon, USA   Bristol, UK
December 2005

Contents

1

Introduction

1.1 Trajectories and Transport in Geophysical Fluid Dynamics

In geophysical fluid dynamics, the evolution and circulation of the ocean and atmosphere are studied as hydrodynamic and thermodynamic phenomena. In many circumstances, molecular diffusion of heat, momentum, or matter is sufficiently weak that the fluid motion of water or air parcels is responsible for most of the transport of these quantities from one location to another. This transport of properties by the bulk motion of fluid is called *advective transport* or *advection*, to distinguish it from the *diffusive transport* accomplished by molecular diffusion. The bulk motion of fluid is itself often called *Lagrangian* motion, to distinguish it from the *Eulerian* perspective, in which the velocity field is analyzed as a function of fixed space and time coordinates, and the paths of fluid parcels are not explicitly considered.

The recognition of the relative importance of advective transport in the mid-latitude atmospheric troposphere was a fundamental insight in the development of numerical weather prediction. Similarly, progress in understanding the ocean's main thermocline has followed the observation that beneath the surface boundary layer, diffusive effects may be small relative to advective effects on time scales of many years or even decades. Although advection is often dominant in geophysical flows, there is an important interplay between advection and diffusion: it is well known that advection can enhance mean diffusion rates, by markedly increasing the magnitudes of small-scale gradients, as in turbulent flows.

Remarkably, however, and despite the dynamical importance of advective transport, the theoretical structure of fluid dynamics is such that it is natural to formulate and solve complete mathematical models of fluid flows, and thus to think quantitatively about fluid motion in general, without determining the trajectories of individual fluid parcels. Fluid dynamics differs in this way from other classical mechanical systems such as particle mechanics, in which it is necessary to keep track of the motion of specific elements of matter in order to develop a workable dynamical description. In a variational framework, this property turns out to be associated with certain symmetries in the fluid Hamiltonian or Lagrangian, and is related to the existence of a materially conserved vorticity quantity.

Because of this structure, the description and dynamical analysis of fluid flows and the description and analysis of the associated trajectories of individual fluid parcels are less tightly linked than one might otherwise anticipate. In numerical modeling, the situation often arises that the velocity field of a flow is known, while no direct information on parcel trajectories exists. This happens also when a velocity field is estimated from measurements of a natural or laboratory fluid flow at certain fixed locations, which is often technically convenient. Alternatively, parcel trajectories may be known while the velocity field is not, as when ocean current measurements are made with drogued drifters or floats. In either case, the motions of fluid parcels may always be regarded as the integral curves of a velocity field. If the velocity field is known by some means, the parcel motion may be obtained by solving a set of ordinary differential equations. Conversely, if a parcel trajectory is known, tangent elements of the velocity field may be obtained by differentiation. It is these aspects—parcel motion and its relation to an associated velocity field—of geophysical fluid dynamics that this text addresses. Specifically, it approaches the problem of quantifying and analyzing advective transport in geophysical fluids from the point of view of dynamical systems theory.

This point of view, which focuses on the relations between the velocity field and the trajectories of fluid elements, purposely avoids most questions regarding the fluid-dynamical origin of the velocity field. Much of fluid dynamics concerns itself with this latter problem, but this text takes an opposite approach. Much of the development of dynamical systems theory over the last century has been stimulated by the deceptively simple general problem of determining and describing the trajectories associated with a given velocity field. In fluid dynamics, this problem has often been given relatively less attention. The present text is partly motivated by the conviction that careful study of this problem still offers many surprises and insights to the physical scientist.

Dynamical systems theory has its roots in the attempt to understand the qualitative behavior of mathematical models of physical systems by studying their long-time asymptotic properties. Typically, in dynamical systems theory, the motions under study are the trajectories of phase-space points that describe the state of the system and its evolution. Here, it is the trajectories of fluid parcels in physical space that is the subject of interest. In either case, the qualitative behavior of different trajectories can be very different: some may circulate endlessly in a small region, while others may rapidly leave the domain of interest, or eventually pass through the vicinity of every point in a large region. For fluid flows, these various behaviors of trajectories have important consequences for the advective transport of fluid properties.

In the ocean, a variety of measurement techniques have revealed the presence of numerous localized, coherent structures at many different scales:

major currents like the *Gulf Stream*, which may extend for thousands of kilometers; mesoscale phenomena including rings, eddies, filaments, and jets, with scales up to hundreds of kilometers; and smaller features such as submesoscale coherent vortices, which may be no more than ten kilometers in width. The atmosphere offers many similar examples, including the *polar vortex*, which circles the globe between the poles and mid-latitudes, and exerts an important and incompletely understood influence on the distribution of ozone in the stratosphere. In many of these flows, the transport and exchange of fluid, and of properties and materials carried by the fluid, between persistent, organized structures or regimes in the flow is of central scientific and practical importance. The term *Lagrangian transport* refers in this text specifically to this advective exchange of fluid parcels between different flow regimes, that is, fluid transport and exchange that is measured with respect to coherent structures in the flow field itself. It is distinguished from traditional Eulerian measures of transport, in which fluxes are computed with respect to a fixed external coordinate system.

The theoretical tools of dynamical systems theory, which were originally developed to address the relations between localized structures and organized motion over extended regions of phase space, offer a natural mathematical framework in which to study Lagrangian transport in these flows. They provide both efficient conceptual descriptions of the field of trajectories and quantitative techniques for computing transports and exchange between different fluid regimes. This approach, which owes its development over the past decade and a half to the efforts of many researchers (see, for example, the references in Section 1.8), makes available many new methods for the study of Lagrangian motion in geophysical fluid flows. In dynamical systems theory, certain characteristic geometric structures appear in phase space, the space of generalized coordinates describing the state of the system. The analogy between fluid flow in physical space and the motion of state vectors in abstract phase spaces motivates the use of these methods to identify and describe the corresponding geometric structures that arise in the context of fluid flow. Trajectories of fluid parcels in certain flows can exhibit a variety of different behaviors, depending on the initial position of the parcel. These different behaviors can often be understood in terms of the corresponding geometric structures that are revealed by dynamical systems analysis of the flow field.

The purpose of this text is to explore some of the new insights into Lagrangian motion in geophysical flows that have followed from this approach, and to present an accessible introduction to the basic elements of some of these methods. These methods are primarily appropriate for strongly inhomogeneous flows with long-lived coherent structures, and the focus here is on a particular set of such flows with special geophysical significance: the closely related phenomena of meandering jets and traveling waves. The next sections introduce the basic fluid-dynamical setting and an analytical model that will be used and explored throughout the text.

1.2 Incompressible Two-Dimensional Flow

Many large-scale geophysical fluid flows are approximately *incompressible* and two-dimensional. Vertical variations in the horizontal flow may be important, but vertical velocities are generally small relative to horizontal velocities, so that even in three-dimensional large-scale flows, incompressible two-dimensional models can offer useful representations of the velocity field at fixed levels or on quasihorizontal surfaces. The jet and wave flows that are our main focus in this text are both of this type, and the models that we discuss all describe motion in only two spatial dimensions. Analysis of Lagrangian motion in three dimensions is an important, but more difficult, problem, and an object of current research.

The Lagrangian motion of fluid elements in two-dimensional flow is represented by their *trajectories*

$$\mathbf{x} = \hat{\mathbf{x}}(\tau; \mathbf{a}) = [\hat{x}(\tau; a, b), \hat{y}(\tau; a, b)], \qquad \hat{\mathbf{x}}(0; \mathbf{a}) = \mathbf{a}, \qquad (1.1)$$

which give, as a function of time τ, the position $\mathbf{x} = (x, y)$ of the element initially (that is, at $\tau = 0$) located at $\mathbf{a} = (a, b)$. Since material fluid elements remain distinct, the fluid motion must be invertible, and corresponding functions may be defined that give the initial position of the element located at $\mathbf{x}$ at time t,

$$\mathbf{a} = \hat{\mathbf{a}}(t; \mathbf{x}) = [\hat{a}(t; x, y), \hat{b}(t; x, y)], \qquad \hat{\mathbf{a}}(0; \mathbf{x}) = \mathbf{x}, \qquad (1.2)$$

where the definition of the inverse means

$$\hat{\mathbf{a}}[t; \hat{\mathbf{x}}(\tau; \mathbf{a})] = \mathbf{a} \quad \text{and} \quad \hat{\mathbf{x}}[\tau; \hat{\mathbf{a}}(t; \mathbf{x})] = \mathbf{x} \quad \text{if} \quad \tau = t. \qquad (1.3)$$

Thus, the *Lagrangian coordinates* $(\mathbf{a}, \tau)$ label fluid elements, while the *Eulerian coordinates* $(\mathbf{x}, t)$ label fixed positions in space. Time has the same meaning in both systems, but τ and t are used to identify the corresponding choices of spatial coordinate system. This identification is particularly helpful when partial derivatives are to be taken with respect to time, since the meaning of these partial derivatives depends critically on whether the Lagrangian or Eulerian spatial coordinates are held fixed when the derivatives are taken. Nonetheless, in a traditional confusion of notation that is continued here, the coordinates $\mathbf{x}$ and $\mathbf{a}$ and their equivalents will often be used also to represent the functions $\hat{\mathbf{x}}$ and $\hat{\mathbf{a}}$ and their equivalents, and t will be often be used to represent time in both the Lagrangian coordinates $(\mathbf{a}, t)$ and the Eulerian coordinates $(\mathbf{x}, t)$. In each case, the meaning should be clear from the context.

The fluid velocity $\dot{\mathbf{x}}$ at each point along a trajectory is the rate of change of position of the corresponding fluid element,

$$\dot{\mathbf{x}} = (\dot{x}, \dot{y}) = \frac{\partial \mathbf{x}(\tau, \mathbf{a})}{\partial \tau}\Big|_{\tau=t} = \left(\frac{\partial x(t; a, b)}{\partial t}, \frac{\partial y(t; a, b)}{\partial t} \right). \quad (1.4)$$

In (1.4), the partial derivatives are taken with the Lagrangian coordinates $\mathbf{a} = (a, b)$ held constant, that is, with respect to τ in the original definitions (1.1), and following the fluid motion. This *material derivative* is often denoted by D/Dt or d/dt in fluid mechanics, so that

$$\dot{\mathbf{x}} = \frac{D\mathbf{x}}{Dt} = \frac{d\mathbf{x}}{dt}. \quad (1.5)$$

The fluid trajectories $\mathbf{x}(t, \mathbf{a})$ are related to the velocity field $\mathbf{v}(\mathbf{x}, t)$ by

$$\dot{\mathbf{x}} = \mathbf{v}(\mathbf{x}, t), \quad (1.6)$$

where the velocity fields for two-dimensional flow have the general form

$$\mathbf{v}(\mathbf{x}, t) = [u(x, y, t), v(x, y, t)]. \quad (1.7)$$

The equations (1.6) are a pair of coupled *ordinary differential equations*, the solutions of which are the trajectories $\mathbf{x}(t; \mathbf{a})$. For a given velocity field $\mathbf{v}(\mathbf{x}, t)$, (1.6) determines $\mathbf{x}(t)$ for each $\mathbf{a}$. In this text, the velocity field $\mathbf{v}(\mathbf{x}, t)$ will always be a known function, and the focus will be entirely on analyzing solutions of (1.6).

Flows that are *incompressible* satisfy the additional condition that the divergence of the velocity field vanishes,

$$\nabla \cdot \mathbf{v} = 0, \quad (1.8)$$

where in the two-dimensional case,

$$\nabla \cdot \mathbf{v} = \frac{\partial u}{\partial x} + \frac{\partial v}{\partial y}. \quad (1.9)$$

This corresponding condition in Lagrangian form is

$$\det J = 1, \quad (1.10)$$

where $\det J$ is the determinant of the Jacobian matrix J,

$$J = \begin{pmatrix} \dfrac{\partial x}{\partial a} & \dfrac{\partial x}{\partial b} \\ \dfrac{\partial y}{\partial a} & \dfrac{\partial y}{\partial b} \end{pmatrix}. \quad (1.11)$$

The constraint (1.8) or (1.10) follows from the principle of mass conservation and the requirement that the fluid density either be uniform or remain constant along trajectories. In the geophysical context, (1.8) is also satisfied by any horizontal *geostrophic* flow on scales small relative to the Earth's radius. For two-dimensional flow, incompressibility implies that the area of a given region of fluid remains constant as the flow evolves (see Appendix A).

A simple but important type of fluid flow is *steady* flow, in which the velocity field $\mathbf{v}$ is independent of time. In this case, the equations for the trajectories are *autonomous*,

$$\dot{\mathbf{x}} = \mathbf{v}(\mathbf{x}). \tag{1.12}$$

For steady flow, all trajectories that pass through a given point are essentially equivalent, differing only by their offsets in time. Consequently, for steady flows the explicit dependence of $\mathbf{x}$ on the initial conditions $\mathbf{a}$ may be dropped, to simplify the notation.

A simple example may be useful to clarify the notation. Consider the steady velocity field $\mathbf{v} = (x, -y)$, with trajectory equations

$$\dot{x} = x, \quad \dot{y} = -y. \tag{1.13}$$

This has solution

$$\mathbf{x}(\tau, \mathbf{a}) = (a\,e^{\tau}, b\,e^{-\tau}), \tag{1.14}$$

where τ has been written for t to emphasize the Lagrangian description. The partial derivative of the solution $\mathbf{x}$ with respect to τ in the Lagrangian coordinate frame is the material derivative,

$$\frac{\partial \mathbf{x}(\tau, \mathbf{a})}{\partial \tau} = (a\,e^{\tau}, -b\,e^{-\tau}) = (x, -y) = \mathbf{v}(\mathbf{x}, t) = \dot{\mathbf{x}}, \tag{1.15}$$

while the partial derivative of the coordinate $\mathbf{x}$ with respect to t in the Eulerian coordinate frame always vanishes identically,

$$\frac{\partial \mathbf{x}}{\partial t} = 0. \tag{1.16}$$

Similarly, for this steady flow example, $\partial \mathbf{v}/\partial t = 0$. This illustrates that if t is used to represent time in the Lagrangian coordinates, and (1.14) is written in the form

$$\mathbf{x}(t, \mathbf{x}_0) = (x_0\,e^{t}, y_0\,e^{-t}), \tag{1.17}$$

care must be taken to preserve the correct meaning of the corresponding partial derivatives. For this example, the inverse function (1.2) that retrieves the Lagrangian label $\mathbf{a}$ from the Eulerian coordinates is

$$\mathbf{a}(t; \mathbf{x}) = (x\,e^{-t}, y\,e^{t}). \tag{1.18}$$

1.3   The Streamfunction

An incompressible two-dimensional velocity field can be expressed entirely in terms of a scalar-valued function known as the *streamfunction*. This is a familiar result in fluid dynamics, but a derivation is provided here for completeness.

From (1.7), (1.9), and *Stokes's theorem*, it follows that

$$0 = - \int \int_{\mathcal{R}} \left(\frac{\partial u}{\partial x} + \frac{\partial v}{\partial y} \right) dx \, dy$$

$$= - \oint_{\mathcal{C}} \mathbf{v} \cdot \mathbf{n} \, ds$$

$$= \oint_{\mathcal{C}} v \, dx - u \, dy, \tag{1.19}$$

where the contour $\mathcal{C}$ is the boundary of the region $\mathcal{R}$, $\mathbf{n}$ is the unit outward normal, ds the differential arc length along $\mathcal{C}$, and the integral is computed counterclockwise around $\mathcal{C}$. If the velocity field $\mathbf{v}$ is time-dependent, the time t is treated as a parameter here, and integration and differentiation are carried out only with respect to the spatial variables x and y, with t fixed. Then, the integrand of the boundary integral on the last line of (1.19) can be represented as the differential of a scalar-valued function of x and y:

$$d\psi(x, y; t) = v(x, y; t) \, dx - u(x, y; t) \, dy. \tag{1.20}$$

On the other hand, the differential of a function $\psi(x, y; t)$ is

$$d\psi(x, y; t) = \frac{\partial \psi}{\partial x}(x, y; t) \, dx + \frac{\partial \psi}{\partial y}(x, y; t) \, dy. \tag{1.21}$$

Since x and y are independent coordinates, the coefficients of the dx and dy terms in (1.20) and (1.21) can be equated to obtain

$$u(x, y; t) = -\frac{\partial \psi}{\partial y}(x, y; t), \qquad v(x, y; t) = \frac{\partial \psi}{\partial x}(x, y; t). \tag{1.22}$$

and the equations for the trajectories are therefore given by

$$\dot{x} = -\frac{\partial \psi}{\partial y}(x, y; t), \qquad \dot{y} = \frac{\partial \psi}{\partial x}(x, y; t), \tag{1.23}$$

where (using (1.20)),

$$\psi(x, y; t) = \int v(x, y; t) \, dx - u(x, y; t) \, dy + \psi_0. \tag{1.24}$$

The integral in (1.24) should be viewed as an indefinite integral, with integration constant ψ_0. Note that the sign convention common in the geophysical fluid literature has been chosen for the streamfunction ψ. In the geophysical context, the streamfunction is often proportional to the dynamic pressure.

Suppose $\mathbf{x}_s(t) = [x_s(t), y_s(t)]$ is a trajectory of a steady incompressible two-dimensional flow with streamfunction $\psi(x, y)$. Then

$$\frac{D}{Dt}\psi[\mathbf{x}_s(t)] = \frac{\partial\psi}{\partial x}\dot{x}_s + \frac{\partial\psi}{\partial y}\dot{y}_s = -\frac{\partial\psi}{\partial x}\frac{\partial\psi}{\partial y} + \frac{\partial\psi}{\partial y}\frac{\partial\psi}{\partial x} = 0, \qquad (1.25)$$

where the partial derivatives of ψ are all evaluated on $\mathbf{x}_s(t)$. Thus, the trajectories lie along the level sets of the streamfunction, that is, along the contours (or curves) of constant ψ defined implicitly by the equation

$$\psi(x, y) = C = \text{constant}. \qquad (1.26)$$

The equation (1.25) also implies that, at each point of a contour of constant ψ, the gradient of the streamfunction is orthogonal to the contour, and the motion along trajectories is tangent to the contour.

1.4 Meandering Jets

The Gulf Stream, a persistent oceanic current or *jet* with a nominal width of 100 km and a peak volume flux of order 10^8 m^3 s^{-1}, flows northward along the east coast of North America, and forms the most prominent feature of the circulation of the upper North Atlantic. When the stream leaves the coast near Cape Hatteras and flows eastward into the open ocean, it develops large meanders: sinuous patterns of lateral displacement that propagate along the stream. Some of these eventually pinch off to form rings, or otherwise result in basic changes in local flow patterns, but often the stream retains the continuous, coherent structure of a meandering jet.

Despite this permanence and coherence, evident in satellite images of sea-surface temperature, measurements suggest that the meandering Gulf Stream exchanges a substantial amount of fluid with its surroundings. Bower & Rossby (1989) found that freely drifting subsurface instruments deployed in the current near Cape Hatteras often quickly left it downstream, and sometimes were subsequently entrained by it again. Reconstructions of these exchange events indicated that these drifting floats typically left the stream just upstream of meander crests or troughs, and rejoined it just downstream of these.

In order to gain insight into this process, Bower (1991) studied the fluid trajectories in an idealized model of a meandering jet. This model was purely kinematic: based on general aspects of the observational record, a representative form of the stream function was proposed, and the corresponding trajectories analyzed, without any specific dynamical reasoning or restrictions. This approach places the focus squarely on the relation between velocity field and fluid trajectory, and provides an important example of the manner in which this relation can become the natural subject

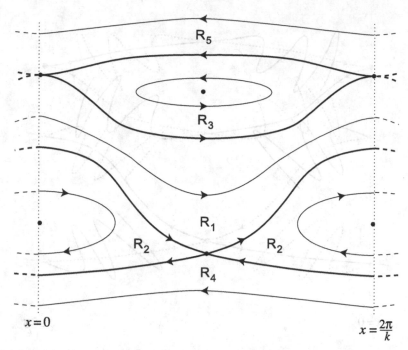

$x = 0$ $\qquad\qquad\qquad\qquad\qquad\qquad\qquad\qquad x = \frac{2\pi}{k}$

FIGURE 1.1. Streamfunction for the kinematic model of a meandering jet, in the frame moving with the meander. The model is periodic in the direction parallel to the mean flow, so that the right and left boundaries of the figure, at $x = 0$ and $x = 2\pi/k$, may be identified. Special level sets of the streamfunction are shown (thick solid lines), which divide the flow into five regimes. The central jet is the region R_1, the closed circulations are R_2 and R_3, and the exterior regions of retrograde motion are R_4 and R_5. A single representative streamline is also shown in each regime (thin solid lines).

of study. The model streamfunction, which will be discussed in detail later on, described a jet of uniform width deformed by a sinusoidal meander that propagated steadily downstream. In a reference frame moving with the meander, the corresponding fluid motion was steady and consisted of three regimes: a central jet, exterior retrograde motion, and intermediate closed circulations above meander troughs and below crests (Figure 1.1).

The motions away from and toward the jet in the closed circulations were originally interpreted as indications of fluid exchange between the jet and its surroundings. This interpretation offered useful insights into the observed behavior of the floats, but the model was incomplete in an important way: strictly speaking, since the fluid motion was steady in the moving reference frame, all parcels remained in their respective regimes, and exchange between regimes did not occur; parcels in the recirculation regime regularly appeared to leave the jet, but always immediately rejoined it at the next trough or crest. This implies that the observed exchange

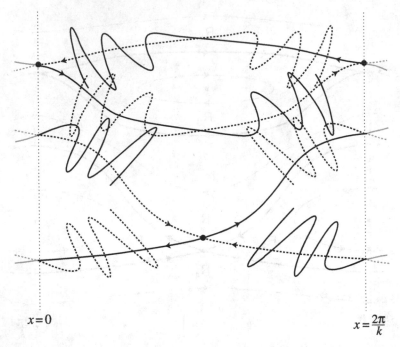

$x=0$

$x=\dfrac{2\pi}{k}$

FIGURE 1.2. Schematic illustration of the breakup of the regime boundaries under the influence of time-dependent disturbances in the moving frame. As in Figure 1.1, the model is periodic in the direction parallel to the mean flow, so that the right and left boundaries of the figure may be identified.

must depend on the presence of additional spatiotemporal variability. If the flow in the moving frame is itself time-dependent, the boundaries between regimes can break up (Figure 1.2), allowing transport and exchange of fluid between regimes.

A central message of this text is that, to analyze and quantify this break-up and the corresponding transport between flow regimes, it is necessary to consider only the special set of solutions that make up the complex tangles of broken steady-flow regime boundaries (Figure 1.2). These trajectories define the boundaries of new, time-dependent flow regimes, describe the geometry of exchange, and are the basic elements of dynamical systems techniques for calculating Lagrangian transport.

1.5   A Kinematic Traveling Wave Model

A slightly simpler example, which is more amenable to analysis, is that of a *traveling wave* in a channel. In a stationary frame of reference (X, Y), the streamfunction for the traveling wave flow is given as

$$\begin{aligned}
\Psi(X, Y, t) &= \Psi_0(X, Y, t) + \varepsilon \Psi_1(X, Y, t) \\
&= A \sin k(X - ct) \sin Y + \varepsilon \Psi_1(X, Y, t)
\end{aligned} \tag{1.27}$$

where $A > 0$ is the amplitude, k is the x-wavenumber, and c, which determines the speed of propagation of the sinusoidal wave-form, is the *phase speed* of the primary wave (see, for example, Knobloch & Weiss (1987), Weiss & Knobloch (1989), and Pierrehumbert (1991a)). The parameter $\varepsilon > 0$ is the amplitude of a second disturbance, whose detailed form is described later. The flow is confined in a channel with rigid boundaries at $Y = 0$ and $Y = \pi$, and the corresponding nonnormal flow conditions are satisfied if $\Psi_1(X, Y = 0, t) = \Psi_1(X, Y = \pi, t) = 0$.

In a reference frame (x, y) moving with the primary wave, with coordinates given by

$$x = X - ct, \quad y = Y, \tag{1.28}$$

the streamfunction $\psi(x, y, t) = \Psi(X, Y, t)$ takes the form

$$\begin{aligned}
\psi(x, y, t) &= \psi_0(x, y) + \varepsilon \psi_1(x, y, t) \\
&= -cy + A \sin kx \sin y + \varepsilon \psi_1(x, y, t).
\end{aligned} \tag{1.29}$$

The equations for the fluid trajectories in the comoving (x, y) frame are

$$\dot{x} = c - A \sin kx \cos y - \varepsilon \frac{\partial \psi_1}{\partial y}(x, y, t)$$

$$\dot{y} = Ak \cos kx \sin y + \varepsilon \frac{\partial \psi_1}{\partial x}(x, y, t). \tag{1.30}$$

This represents the superposition of a time-dependent disturbance $\varepsilon \psi_1$ on a two-dimensional steady velocity field in the comoving frame.

For $\varepsilon = 0$ and c chosen appropriately, (1.27) is an exact solution to the *barotropic vorticity* (or *potential vorticity*) equation (see, for example, Pierrehumbert (1991a)), in a channel with rigid boundaries at $y = 0$ and $y = \pi$. For $A = 1$, $k = 1$, $c = 0.5$, and $\varepsilon = 0$, the flow is periodic with x-wavelength 2π (Figure 1.3). The structure of the flow in the comoving frame (x, y) is similar to that of the meandering jet described above (Figure 1.1): there is a contiguous jet flowing in the x-direction, a closed clockwise recirculation adjacent to the wall $y = 0$, and a closed counterclockwise circulation adjacent to the wall $y = \pi$. These jet and recirculation regions are separated by bounding streamlines (Figure 1.3), and there is no transport or exchange between them. In contrast to the meandering jet, the traveling wave is bounded laterally by rigid channel walls, and lacks an external region of retrograde motion.

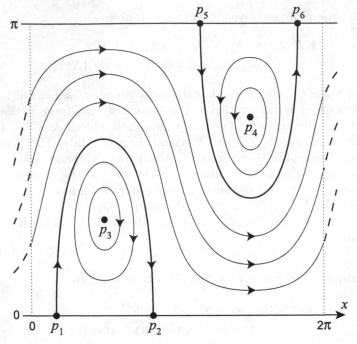

FIGURE 1.3. Steady streamlines in the comoving frame for the traveling wave flow (1.29), for $A = 1$, $k = 1$, $c = 0.5$, and $\varepsilon = 0$. The channel is bounded by rigid walls at $y = 0$ and $y = \pi$. There are three flow regimes, separated by the two special streamlines (thick solid lines) that join pairs of stagnation points, (p_1, p_2) and (p_5, p_6): two recirculation regions around the interior stagnation points p_3 and p_4, and a single, central jet region between the recirculations. Only the segment $0 < x < 2\pi$ is shown, corresponding to one wavelength of the traveling wave.

1.6   Critical Lines and Cellular Flow

A characteristic feature of simple meandering-jet and traveling-wave flows is the presence, in the comoving frame, of regions of recirculating fluid along the sides of a contiguous central jet (Figures 1.1 and 1.3). All the fluid in these regions travels at a mean speed equal to the phase speed of the meander or the traveling wave. Such regions of trapped, comoving fluid are found in many types of nonlinear wave flows.

Related features often also appear when small-amplitude waves propagate into a jet from a quiescent region. In that case, the analogous recirculation cells will form along the lines where the speed of propagation (that is, the phase speed) of the incident wave is equal to the undisturbed jet speed, and these lines are called *critical lines*. The asymptotically narrow recirculation cells that develop in the *critical layers* that surround these lines are referred to as *cat's eyes*, because of their shape and the manner

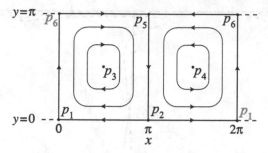

FIGURE 1.4. Streamlines for the steady, cellular flow. A single along-channel wavelength of the periodic flow is shown. In this case, there are only two flow regimes. These are both recirculations, separated by the special cross-channel streamlines (thick solid lines) that join the stagnation point pairs (p_1, p_6) and (p_2, p_5), and there is no jet regime.

in which they open and close proportionally with the amplitudes of the incident waves. This term is sometimes used more generally to refer also to the recirculation cells that are present in the large-amplitude, nonlinear case.

The size and structure of the recirculation cells in meandering-jet and traveling-wave flows are influenced by many factors, but, like the cat's eyes of small-amplitude theory, often depend importantly on the ratio of the phase speed of the meander or traveling wave to the speed of the fluid parcel motion in the stationary frame. When the phase speed is larger than the speed of the fluid parcel motion, such regions typically do not form. For example, a jet with small-amplitude meanders will in this case have no critical lines, since there will be no points where the phase speed and jet speed match. As the phase speed decreases, recirculation cells generally form and grow. Eventually, they may expand so far that the contiguous central jet is entirely expunged and the counterrotating cells on opposite sides are in direct contact. This process is sometimes referred to as *reconnection* (Bower & Rossby (1989), Meyers (1994), Pratt et al. (1995)). In the case of the simple *unperturbed* traveling wave, (1.1) with $\varepsilon = 0$, this *cellular flow* limit is reached when the phase speed c vanishes, so that the streamfunction in the traveling frame is simply $\psi(x, y, t) = A \sin kx \sin y$ (Figure 1.4).

1.7 The Onset of Fluid Exchange

When the two-dimensional flow in the translating frame is steady, the traveling wave contains several distinct flow regimes, separated by bounding streamlines, with no transport or exchange between them (Figures 1.3, 1.4). This picture will change if there is additional time-dependence in the traveling-wave velocity field, that is, if $\varepsilon \neq 0$ in (1.27) and (1.29).

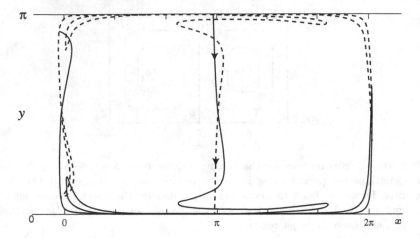

FIGURE 1.5. Breakup of regime boundaries for the cellular flow with a sinusoidal, time-periodic perturbation to the steady streamfunction.

Analytical, numerical, and laboratory studies, along with general considerations from dynamical systems theory, lead us to anticipate that, when an additional time-dependent motion is present, these cell boundaries typically break up, giving rise to Lagrangian motions that exchange fluid between cells in wildly oscillating lobe structures along the edges of the cells (Figure 1.5). Thus, the dominant mode of large-scale fluid exchange and mixing associated with meandering jets and traveling waves of moderate amplitude involves the stirring action of coherent, fluctuating centers of recirculation and their interaction with each other and with the shear flow in which they are embedded. It is the analysis and description of complex Lagrangian motions of this type that we explore in this text. An important goal of the following chapters is to explain clearly the meaning of illustrations such as Figures 1.2 and 1.5, and the structures represented in them.

1.8   Notes

The dynamical systems approach to fluid transport and exchange that is described in this text was primarily developed in the 1980s and 1990s. Much of the early motivation for this work came from engineering problems, such as the design of efficient mixing systems for viscous fluids. This history is described by Aref (2002). Ottino (1989) provides a useful, complementary introduction that is oriented toward the engineering perspective. Other general references include Wiggins (1992) and the conference proceedings Acrivos et al. (1991) and Babiano et al. (1994).

Many of the geophysical fluid analyses that use these ideas have focused on transport and exchange in meandering jet and traveling wave flows. The early work of Regier & Stommel (1979) partially motivated the Bower (1991) and Samelson (1992) models of Lagrangian motion in the Gulf Stream that are discussed above. The traveling-wave example (1.27) is taken from the atmospherically motivated analysis by Pierrehumbert (1991a). Similar ideas were developed by Knobloch & Weiss (1987). Subsequent extensions include the studies by Dutkiewicz et al. (1993), Meyers (1994), Duan & Wiggins (1996), and Cencini et al. (1999). Motivated by the laboratory experiments of Sommeria et al. (1989), del Castillo-Negrete & Morrison (1993) independently developed traveling-wave models and analyzed the associated Lagrangian transport. These studies generally relied on kinematic or linearized descriptions of the velocity fields. Some recent work that has begun to extend this perspective to flows described by dynamical models is discussed in Chapter 7 and in the recent review Wiggins (2005). Lagrangian transport in the idealized but geophysically relevant case of laboratory thermal convection has been studied theoretically and in the laboratory by Clever & Busse (1974), Bolton et al. (1986), Camassa & Wiggins (1991), and Solomon & Gollub (1988). Useful general references for the fundamental elements of the theory of geophysical fluid dynamics include Pedlosky (1987), Salmon (1998), and Holton (1992). Of particular interest for its Lagrangian perspective is the recent text on Lagrangian fluid dynamics by Bennett (2006).

2

Steadily Translating Waves and Meanders

2.1 The Comoving Frame

A common element of the simplest models of meandering jets and traveling waves is that in their most basic form, they are characterized by the steady propagation of a meander or wave pattern at a single, constant phase speed. In a frame of reference moving at this same speed, the corresponding velocity fields are then steady. The analysis of the associated Lagrangian motion is considerably simplified if it is conducted in the comoving frame. For the traveling-wave flow (1.27), a steady streamfunction in the comoving frame results when the disturbance amplitude ε in (1.29) is equal to zero, so that $\psi(x, y, t) = \psi_0(x, y)$ and

$$\dot{x} = -\frac{\partial \psi_0}{\partial y} = c - A \sin kx \, \cos y,$$

$$\dot{y} = \frac{\partial \psi_0}{\partial x} = Ak \cos kx \, \sin y, \tag{2.1}$$

are the corresponding autonomous equations for the fluid trajectories.

Some aspects of the resulting picture have been briefly illustrated in Chapter 1. Indeed, since the streamfunction is known, a complete characterization of the motion can be easily achieved in this specific case, simply by considering the structure of the level sets of the streamfunction. In preparation for the analysis of the unsteady case, however, in which the velocity field must be considered directly, it is useful to proceed here with a review of basic elements of motion in steady, two-dimensional, incompressible flows. Thus, in this chapter we consider steady velocity fields of the form

$$\mathbf{v}(\mathbf{x}) = [u(x, y), v(x, y)], \tag{2.2}$$

the trajectories of which are the solutions of the autonomous equations (1.12). We use (2.1) as an example, and use general methods to analyze the steady flow in the translating frame for the traveling wave. The main subject of the following chapters will be the analysis of the complexities that arise when this steady motion in the translating frame is perturbed by the introduction of additional disturbances, or, generally, of more complex time-dependence in the jet and wave flows.

2.2   Stagnation Points

The simplest fluid trajectories in steady flows are the stagnation points. A
stagnation point of the velocity field $\mathbf{v}(\mathbf{x})$ is a point $\mathbf{x}_0 = (x_0, y_0)$ satisfying

$$\mathbf{v}(\mathbf{x}_0) = 0. \tag{2.3}$$

Since a fluid element located at a stagnation point does not move, its
trajectory consists of the single point $\mathbf{x}_0$, the stagnation point itself.

For the traveling-wave flow in the translating frame with $\varepsilon = 0$, there
are stagnation points if $|c| < A$ (Figure 1.3). The three pairs of stagnation
points (x_0, y_0) in $0 \le x \le 2\pi/k$ can be easily obtained from (2.1). Along
the channel boundaries, there are two pairs, each of which satisfies one of
the two equations

$$\sin k x_0 = \frac{c}{A}, \quad y_0 = 0, \tag{2.4}$$

and

$$\sin k x_0 = -\frac{c}{A}, \quad y_0 = \pi. \tag{2.5}$$

In the interior of the flow, there are two additional stagnation points, which
satisfy

$$x_0 = \frac{\pi}{2k}, \quad \cos y_0 = \frac{c}{A}, \tag{2.6}$$

and

$$x_0 = \frac{3\pi}{2k}, \quad \cos y_0 = -\frac{c}{A}. \tag{2.7}$$

Label the stagnation points $p_1, \ldots, p_6$, consecutively with increasing x in
$0 \le x \le 2\pi/k$, and let $x_s = \sin^{-1}(c/A)$, and $y_s = \cos^{-1}(c/A)$. Then
$p_1 = (x_s/k, 0)$, $p_2 = ((\pi - x_s)/k, 0)$, $p_3 = (\pi/2k, y_s)$, $p_4 = (3\pi/2k, \pi -
y_s)$, $p_5 = ((\pi + x_s)/k, \pi)$, and $p_6 = ((2\pi - x_s)/k, \pi)$. For the parameter
values $A = k = 1$, $c = 0.5$, we have (Figure 1.3) $p_1 = \left(\frac{\pi}{6}, 0\right)$, $p_2 = \left(\frac{5\pi}{6}, 0\right)$,
$p_3 = \left(\frac{\pi}{2}, \frac{\pi}{3}\right)$, $p_4 = \left(\frac{3\pi}{2}, \frac{2\pi}{3}\right)$, $p_5 = \left(\frac{7\pi}{6}, \pi\right)$, and $p_6 = \left(\frac{11\pi}{6}, \pi\right)$. Identical
repetitions of these three pairs of stagnation points occur in each section
$2\pi j \le x \le 2\pi(j + 1)$, $j = \{\ldots, -1, 0, 1, \ldots\}$ of the channel.

For $0 < c < A$, the stagnation points that lie along the boundaries also
lie on streamfunction contours that separate the recirculation cells from
the contiguous central jet (Figure 1.3). Analogous stagnation points lie
on streamfunction contours that separate the recirculation cells (R_2 and
R_4) from the jet (R_3) and exterior (R_1 and R_5) regions in the meandering jet flow (Figure 1.1). These contours and stagnation points thus have
special significance, since they define a set of regions in each of which
the Lagrangian motion of different fluid parcels is qualitatively similar.
In contrast, the stagnation points in the centers of the recirculation cells
(Figures 1.1 and 1.3) are surrounded by fluid-parcel trajectories that simply encircle them. We will return below to this distinction, and to a more

detailed consideration of the trajectories on the boundaries of the recirculation cells, after conducting a quantitative analysis of the Lagrangian motion in the vicinity of the stagnation points.

2.3 Linearization near Stagnation Points

Stagnation points are of special interest because, in addition to the singular character of the associated trajectories, they are points where the flow characteristics often change. Along any line through a stagnation point, the flow on opposite sides of the stagnation point is typically in opposite directions. Here, we consider in detail the behavior of the fluid trajectories near stagnation points. We review the solution of the general problem first, and then return to the analysis of the traveling-wave stagnation points.

In order to understand the behavior of trajectories near a stagnation point, we must consider the structure of the velocity field near the stagnation point. For a given stagnation point $\mathbf{x}_0 = (x_0, y_0)$, we make the coordinate transformation $\mathbf{x} = \mathbf{x}' + \mathbf{x}_0$, and then expand the right-hand side of (1.12) in a Taylor series in $\mathbf{x}'$, using (2.3) to obtain

$$\dot{x}' = \frac{\partial u}{\partial x}(\mathbf{x}_0)x' + \frac{\partial u}{\partial y}(\mathbf{x}_0)y' + \mathcal{O}(|\mathbf{x}'|^2),$$

$$\dot{y}' = \frac{\partial v}{\partial x}(\mathbf{x}_0)x' + \frac{\partial v}{\partial y}(\mathbf{x}_0)y' + \mathcal{O}(|\mathbf{x}'|^2). \tag{2.8}$$

The notation $\mathcal{O}(|\mathbf{x}'|^2)$ represents terms of quadratic and higher order in the components of $\mathbf{x}'$. The linear approximation to, or *linearization* of, (2.8) is obtained by neglecting these quadratic and higher-order terms.

Written in matrix form, the linearization of the velocity field $\mathbf{v}$ at the stagnation point $\mathbf{x}_0$ is

$$\begin{pmatrix} \dot{x}' \\ \dot{y}' \end{pmatrix} = \begin{pmatrix} \dfrac{\partial u}{\partial x}(\mathbf{x}_0) & \dfrac{\partial u}{\partial y}(\mathbf{x}_0) \\ \dfrac{\partial v}{\partial x}(\mathbf{x}_0) & \dfrac{\partial v}{\partial y}(\mathbf{x}_0) \end{pmatrix} \begin{pmatrix} x' \\ y' \end{pmatrix}, \tag{2.9}$$

where the partial derivatives are evaluated at $\mathbf{x} = \mathbf{x}_0$. In matrix notation, (2.9) may be written,

$$\dot{\mathbf{x}}' = D\mathbf{v}(\mathbf{x}_0)\mathbf{x}'. \tag{2.10}$$

Here $D\mathbf{v}(\mathbf{x}_0)$ is referred to as the *velocity gradient tensor*, the *Jacobian matrix*, or the *matrix associated with the linearization*. Note that, by linearity, $D\mathbf{v}(\mathbf{x}_0)(-\mathbf{x}') = -D\mathbf{v}(\mathbf{x}_0)\mathbf{x}'$. This explains the flow reversal about any line through a stagnation point mentioned above.

2.4   Trajectories of Linearizations

The trajectories of the linearization may be obtained by solving the linear differential equations (2.10). For two-dimensional, incompressible flow, these trajectories of the linearization correctly describe the qualitative behavior of trajectories of the full nonlinear velocity field in a sufficiently small region near the stagnation point. General characteristics of this motion field can thus be inferred from general properties of solutions of equations of the form (2.10).

It is well known that the relative motion near a point can be analyzed in terms of shear, strain, and rotation (e.g., Batchelor (1967)). Each of these categories of motion corresponds to a particular eigenvalue and eigenspace structure of the matrix $D\mathbf{v}(\mathbf{x}_0)$, which determines the qualitative structure of the trajectories for the linearized velocity field.

Since the sum of the eigenvalues of $D\mathbf{v}(\mathbf{x}_0)$ must equal the trace of the matrix, which is the divergence of the velocity field at the stagnation point, this sum must be zero if the fluid is incompressible. Thus, for two-dimensional, incompressible flows, there are only three possibilities for the pair of eigenvalues of the velocity gradient tensor:

$$\text{I (shear/no flow)}: 0,0, \quad \text{II (rotation)}: \pm i\omega, \quad \text{III (strain)}: \pm\lambda, \quad (2.11)$$

where $\omega, \lambda > 0$, and $i := \sqrt{-1}$. Case I corresponds either to pure shear or to no flow, case II to pure rotation, and case III to pure strain. By a linear coordinate transformation, matrices with these eigenvalues can be put in the following real Jordan canonical forms:

$$\text{Ia (no flow)}: \begin{pmatrix} 0 & 0 \\ 0 & 0 \end{pmatrix}, \quad \text{Ib (shear)}: \begin{pmatrix} 0 & 1 \\ 0 & 0 \end{pmatrix},$$

$$\text{II (rotation)}: \qquad \begin{pmatrix} 0 & -\omega \\ \omega & 0 \end{pmatrix}, \qquad (2.12)$$

$$\text{III (strain)}: \qquad \begin{pmatrix} \lambda & 0 \\ 0 & -\lambda \end{pmatrix},$$

where there are two possibilities for case I because of the repeated eigenvalue zero. A result that will be familiar to fluid dynamicists is that a shearing motion can be represented as the sum of a pure strain and a pure rotation (e.g., (Batchelor, 1967, Section 2.3)). Perhaps more surprising is that, in suitable coordinates, two-dimensional, incompressible, relative motion around a stagnation point always takes exactly one of these four forms.

If we denote the matrices in (2.12) by M, then the corresponding equation for the trajectories of the linear velocity field can be written

$$\begin{pmatrix} \dot{\xi} \\ \dot{\eta} \end{pmatrix} = M \begin{pmatrix} \xi \\ \eta \end{pmatrix}. \qquad (2.13)$$

Here we have used the notation (ξ, η) for the spatial coordinates, rather than $\mathbf{x}' = (x', y')$ as in (2.10), to emphasize that, in general, a linear transformation of the coordinates is necessary in order to put $D\mathbf{v}(\mathbf{x}_0)$ in one of the forms in (2.12).

For each of the linearized velocity fields (2.13), the equations for the trajectories can be directly solved (Figure 2.1):

Case	Velocity Field	Trajectories	
Ia (no flow) :	$\dot{\xi} = 0$ $\dot{\eta} = 0$	$\xi(t) = \xi_0 = \text{constant}$ $\eta(t) = \eta_0 = \text{constant}$	
Ib (shear) :	$\dot{\xi} = \eta$ $\dot{\eta} = 0$	$\xi(t) = \eta_0 t + \xi_0$ $\eta(t) = \eta_0 = \text{constant}$	(2.14)
II (rotation) :	$\dot{\xi} = -\omega\eta$ $\dot{\eta} = \omega\xi$	$\xi(t) = \xi_0 \cos\omega t - \eta_0 \sin\omega t$ $\eta(t) = \xi_0 \sin\omega t + \eta_0 \cos\omega t$	
III (strain) :	$\dot{\xi} = \lambda\xi$ $\dot{\eta} = -\lambda\eta$	$\xi(t) = e^{\lambda t}\xi_0$ $\eta(t) = e^{-\lambda t}\eta_0$	

For case Ia all trajectories are stagnation points. For case Ib all trajectories are straight lines with $\eta_0 = \text{constant}$. For $\eta_0 > 0$ trajectories move to the right, for $\eta_0 < 0$ trajectories move to the left, and all trajectories with $\eta_0 = 0$ are stagnation points. For case II all trajectories are periodic with frequency ω. For case III, the trajectories are exponential.

Stagnation points for which the eigenvalues of the velocity gradient tensor are real and opposite in sign (pure strain) are *saddle points*. In this case, the fluid trajectories for the linearized flow around the stagnation points are hyperbolas. This can be seen directly from case III of (2.14) by noting that trajectories in this case in the transformed coordinates satisfy $\xi(t)\eta(t) = \text{constant}$. For general three-dimensional flows, the term *hyperbolic* is used to refer to any stagnation point having the property that none of the eigenvalues of the matrix associated with the linearization have zero real part. Hence, for incompressible, two-dimensional flows, the only *hyperbolic stagnation points* are the saddle points.

Stagnation points for which the eigenvalues of the velocity gradient tensor are purely imaginary (the case of pure rotation) are *centers* or *elliptic points*, and the corresponding fluid trajectories for the linearized two-dimensional flow are ellipses in the original coordinate frame. This can be seen directly from (2.14) by noting that trajectories in case II satisfy $\xi(t)^2 + \eta(t)^2 = \text{constant}$ and so are circles in the transformed coordinates.

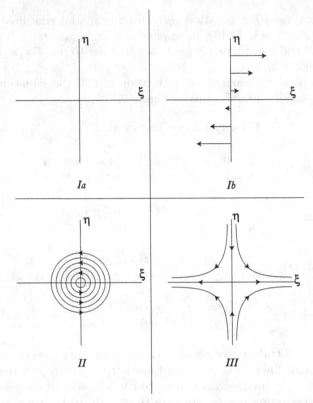

FIGURE 2.1. Trajectories for the four cases for linear, steady, incompressible, two-dimensional velocity fields. In case Ia (no flow) all points are stagnation points. In case Ib (pure shear) all points on the ξ axis are stagnation points. In cases II (pure rotation) and II (pure strain) the only stagnation point is at the origin.

2.5 The Traveling Wave: Linearizations

Having reviewed general results for linearized velocity fields in the last two sections, we now return to the problem of the Lagrangian motion in the traveling wave (1.27) with $\varepsilon = 0$. At a stagnation point $\mathbf{x}_0 = (x_0, y_0)$ of the traveling wave in the translating frame (1.29), the linearized velocity field is

$$Dv(\mathbf{x}_0) = \begin{pmatrix} -Ak\cos kx_0\cos y_0 & A\sin kx_0\sin y_0 \\ -Ak^2\sin kx_0\sin y_0 & Ak\cos kx_0\cos y_0 \end{pmatrix}. \qquad (2.15)$$

The eigenvalues of $Dv(\mathbf{x}_0)$ are given by

$$\lambda_\pm = \pm\sqrt{-\det Dv(\mathbf{x}_0)}, \qquad (2.16)$$

where

$$\det D\mathbf{v}(\mathbf{x}_0) = -A^2 k^2 \left(\cos^2 kx_0 \cos^2 y_0 - \sin^2 kx_0 \sin^2 y_0\right). \qquad (2.17)$$

For the boundary stagnation points (2.4) and (2.5), the second term in parentheses in (2.17) vanishes, and $\det D\mathbf{v}(\mathbf{x}_0)$ is negative. Hence the stagnation points on the boundary are all saddles (and so hyperbolic, in the more general terminology). For the interior stagnation points (2.6) and (2.7), the first term in parentheses in (2.17) vanishes, and $\det D\mathbf{v}(\mathbf{x}_0)$ is positive. Hence the stagnation points in the recirculation cells are all elliptic. These analytical results for the traveling wave are consistent with the streamfunction contours near the stagnation points (Figure 1.3): these contours intersect at the boundary stagnation points, with motion toward and away from those points (as in Figure 2.1, case III), and encircle the interior stagnation points (as in Figure 2.1, case II).

For the traveling wave, (2.17) can be evaluated analytically at each stagnation point. Let $\delta_s = [1 - (c/A)^2]^{1/2}$. Then

$$D\mathbf{v}(p_1) = -D\mathbf{v}(p_2) = D\mathbf{v}(p_5) = -D\mathbf{v}(p_6) = Ak\delta_s \begin{pmatrix} -1 & 0 \\ 0 & 1 \end{pmatrix}, \qquad (2.18)$$

$$D\mathbf{v}(p_3) = -D\mathbf{v}(p_4) = A\delta_s \begin{pmatrix} 0 & 1 \\ -k^2 & 0 \end{pmatrix}. \qquad (2.19)$$

The Jacobian matrices in (2.18) are all diagonal, so in each case there is one eigenvector in the x-direction and one eigenvector in the y-direction, and the sign of the respective eigenvalue gives the direction of motion along the eigenvector near the hyperbolic points p_1, p_2, p_5, and p_6. The antisymmetric Jacobians in (2.19) lead to oscillatory motion around the interior elliptic points p_3 and p_4. For the parameter values $A = k = 1$, $c = 0.5$, we have (Figure 2.2) $\delta_s = \sqrt{3}/2$.

2.6 Material Curves and Invariant Subspaces

A *material curve* is a curve in the fluid domain that is composed of points that move with the fluid. For the general case of pure strain (case III discussed above, the saddle or hyperbolic stagnation point), a special set of linear material curves may be identified, which correspond to the eigenspaces of the linearized velocity field.

Consider any trajectory of case III in (2.14) that starts on the ξ axis, that is, any trajectory for which $\eta_0 = 0$. For any such trajectory, η always remains zero, since $\dot{\eta} = 0$. Thus, the ξ axis is a material curve. The set of all such trajectories with $\eta_0 = 0$ naturally forms a *vector space*. Since this set is also a material line, it is also invariant under the flow, in the sense that points initially on the line remain on the line under the evolution of

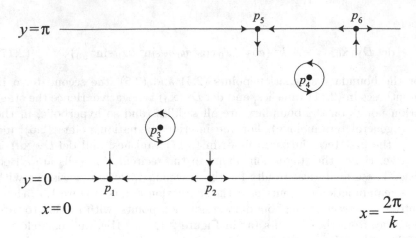

FIGURE 2.2. Trajectories of the velocity field of the traveling wave linearized near the stagnation points.

the flow. The ξ axis is therefore also an *invariant subspace* for the linearized velocity field of case III.

Since the ξ axis is a material line, no trajectory can ever cross it. Thus the ξ axis divides the linearized velocity field into two *flow domains*: the upper half-plane ($\eta > 0$) and the lower half-plane ($\eta < 0$), and no trajectory can pass from one of these two domains into the other. Trajectories on the ξ axis have the property that their distance from the stagnation point grows at an exponential rate, as is apparent from the expression given in (2.14) for a trajectory starting from a point on the ξ axis in case III. Thus, the ξ axis is the *unstable subspace*, denoted by E^u, corresponding to the span of the eigenvector of M associated with the positive eigenvalue λ.

Similarly, the η axis is also a material line and an invariant subspace. It is the *stable subspace*, denoted by E^s, corresponding to the span of the eigenvector of M associated with the negative eigenvalue $-\lambda$. Trajectories starting on the η axis approach the stagnation point at an exponential rate.

Since trajectories cannot cross from one side of either of these invariant subspaces to the other, the stable and unstable subspaces divide the flow locally into regions: trajectories starting in a given region must remain in that region. Thus, near the saddle points (p_1, p_2, p_5, and p_6), trajectories of the unperturbed traveling wave flow belong to distinct regions that may be distinguished by the direction of flow along the channel boundaries (Figure 2.2). In general, the linearization will become inaccurate when the trajectories reach finite distances from the saddle points, but the linearization does give the correct description near the saddle points.

2.7 Material Manifolds of Stagnation Points

A streamline defined by (1.26) is an *invariant manifold*—that is, an invariant curve or surface—of the corresponding flow: if a trajectory starts on the streamline, then it remains on the streamline. Geometrically, this is manifested by the velocity field being tangent to the streamline, as we showed above. As with the invariant linear subspaces considered above, no trajectory can pass through an invariant manifold, from one side to the other, for this would violate the uniqueness property of solutions to (1.12). For this reason, invariant manifolds play a central role in the mathematical description of the transport of matter and material properties in fluid flows, as we will see later on in this book. Physically, invariant manifolds are material curves or surfaces. To emphasize their material character in the fluid mechanical setting, we introduce here the term *material manifold* to describe these objects.

In Section 2.3, we described the trajectories near stagnation points for the linearized velocity field, and concluded that a special set of material lines existed, the invariant subspaces E^u and E^s, that divided the trajectories of the linearized velocity field into different flow regimes. In the context of the linearized velocity field, these subspaces are examples of material manifolds. The *Morse lemma* guarantees that the trajectories of the linearized velocity field will be qualitatively accurate descriptions of the trajectories of the full velocity field in at least a small region near the stagnation point. Thus, if the stagnation point is elliptic by the linear criterion, the Morse lemma implies that it is surrounded by closed trajectories of the full velocity field, and if the stagnation point is a saddle by the linear criterion, the Morse lemma implies that it is also a saddle point with regard to the trajectories of the full velocity field.

In the case of the linearized velocity field for the unperturbed traveling wave, we saw that the material lines E^u and E^s divided the flow near the saddle points (p_1, p_2, p_5, and p_6) into different flow regimes. For incompressible, two-dimensional flows, an argument using the Morse lemma

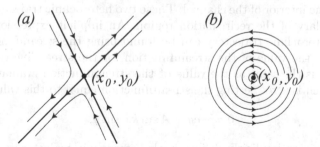

FIGURE 2.3. The geometry of streamlines near (a) hyperbolic (saddle) and (b) elliptic (center) stagnation points.

shows that a similar pair of material curves must pass through the saddle point for the full velocity field, and again divide the flow into different flow regimes (Figure 2.3). These curves may not be straight lines in the original coordinates, because of the nonlinearity. They will, however, be tangent to the invariant subspaces E^s and E^u at the stagnation point, and the associated trajectories will have the corresponding asymptotic properties. This special pair of material curves are called the *stable and unstable manifolds* of the saddle point. The stable manifold consists of all those points through which the corresponding trajectories are asymptotic to the saddle point as $t \to \infty$. Similarly, the unstable manifold consists of all those points through which the corresponding trajectories are asymptotic to the saddle point as $t \to -\infty$. We refer to these curves as *material manifolds* of the saddle point, to emphasize their character as material curves: they are composed of points that move with the fluid.

2.8 The Traveling Wave: Material Manifolds

For the traveling wave in the translating frame (1.29) with $\varepsilon = 0$, substituting $y = 0$ into the velocity field gives $\dot{y} = 0$. Hence, trajectories with $y = 0$ initially have $y = 0$ for all time. Therefore the x axis, which is composed of all the points with $y = 0$, is an invariant manifold. A similar argument, and conclusion, follows for the line $y = \pi$.

Motion does occur along $y = 0$, except at the stagnation points p_1 and p_2. Consideration of the direction of this motion reveals that a trajectory that starts on the x axis between p_1 and p_2 must approach p_1 as $t \to +\infty$, and approach p_2 as $t \to -\infty$. This is an example of a *heteroclinic trajectory*: a trajectory that asymptotically approaches two different stagnation points as $t \to +\infty$ and $t \to -\infty$. (A trajectory that asymptotically approaches the same stagnation point for $t \to +\infty$ and $t \to -\infty$ is called a *homoclinic trajectory*.) A similar argument, and discussion, follows for the segment of $y = \pi/l$ between p_5 and p_6.

There is another heteroclinic trajectory between p_1 and p_2, which passes through the interior of the channel. These two heteroclinic trajectories form the boundary of the recirculation region. An implicit expression for the second heteroclinic trajectory can be found using the streamfunction. By continuity, the value of the streamfunction for the streamline connecting p_1 and p_2 is the same as the value of the streamfunction evaluated on p_1 or p_2, which is zero. Setting the streamfunction equal to this value gives

$$0 = -cy + A \sin kx \sin y. \tag{2.20}$$

This expression implicitly defines both heteroclinic trajectories, and thus the set of possible initial conditions for trajectories that approach either p_1 or p_2 as $t \to \pm\infty$. The second heteroclinic trajectory, with $y \neq 0$, then satisfies

$$x(t) = \frac{1}{k} \sin^{-1} \left[\frac{cy(t)}{A \sin y(t)} \right], \qquad -\infty < t < \infty, \qquad (2.21)$$

which is an implicit expression for $(x(t), y(t))$.

A similar discussion holds for the heteroclinic trajectories connecting p_5 and p_6. The implicit expression that both heteroclinic trajectories must satisfy is given by

$$x(t) = \frac{1}{k} \sin^{-1} \left[-\frac{c(\pi - y(t))}{A \sin y(t)} \right], \qquad -\infty < t < \infty. \qquad (2.22)$$

All of these heteroclinic trajectories are examples of material manifolds. Some general aspects of their structure can be discerned from (2.21) and (2.22). For $|c| < A$, the recirculation zones become larger as c decreases toward zero. At $c = 0$, the jet disappears altogether and we obtain the cellular flow (Figure 1.4).

For the cellular flow, it is straightforward to compute the material manifolds explicitly, and to see how these delineate the different flow regimes. The stagnation points in the middle of each cell are elliptic, so these are surrounded by closed trajectories. The stagnation points on the boundary are saddle points. Consider the cell $0 \le x \le \pi/k$, $0 \le y \le \pi$. The segment $y = 0$, $0 < x < \pi/k$, is a stable manifold of the stagnation point $(x, y) = (0, 0)$ *and* simultaneously an unstable manifold of $(x, y) = (\pi/k, 0)$. A similar situation holds for the upper boundary $y = \pi$. Now consider the saddle-type stagnation points $(x, y) = (0, 0)$ and $(x, y) = (0, \pi)$. Since $\dot{x} = 0$ on the vertical line $x = 0$, $0 < y < \pi$, this vertical line is a material manifold, which is simultaneously the unstable manifold of $(x, y) = (0, 0)$ and the stable manifold of $(x, y) = (0, \pi)$. A similar situation holds for the vertical line $x = \pi/k$, $0 < y < \pi$. Thus, the cell boundaries are composed of hyperbolic stagnation points and their associated material manifolds.

It takes an infinite time ($t \to +\infty$ or $t \to -\infty$) for a trajectory through a point on a stable or unstable material manifold to reach either of the stagnation points at the ends of the manifolds. This can be seen in two ways. First, if either could be reached in finite time, this would violate uniqueness of solutions, since the stagnation point is itself a trajectory, and two different trajectories must remain distinct. Second, the boundary of the cell can be viewed as the limit of periodic trajectories, for which explicit expressions are given in Appendix B. The expressions (B.20) and (B.29) for the periods of the closed trajectories show that the period goes to infinity as the boundary is approached.

2.9 Notes

This chapter reviews standard results on the trajectories of the linearized velocity field at a stagnation point, from a geometrical point of view. Comprehensive discussions of the geometric theory of linear differential equations can be found in Arnold (1973) and Hirsch & Smale (1974). The *stable and unstable manifold theorem for hyperbolic stagnation points* and the related concepts of homoclinic and heteroclinic trajectories are discussed by Arnold (1973) and Hirsch & Smale (1974), and by Wiggins (2003).

3

Integrability of Lagrangian Motion

3.1 Scalar Advection: The Material Derivative

The usual goal of a Lagrangian analysis is to understand the motion and evolution of objects, matter, or properties that are carried along by the flow. The *material derivative* DF/Dt is the rate of change of a scalar property or function $F(x, y, t)$ following the fluid motion:

$$\frac{DF}{Dt} = \frac{\partial F}{\partial \tau} = \dot{F} = \frac{\partial F}{\partial t} + \mathbf{v} \cdot \nabla F, \qquad (3.1)$$

where the symbol τ has been used to denote time in the Lagrangian coordinate system, as in (1.1)–(1.3) For two-dimensional, incompressible flow with streamfunction ψ,

$$\frac{DF}{Dt} = \frac{\partial F}{\partial t} + u \frac{\partial F}{\partial x} + v \frac{\partial F}{\partial y} = \frac{\partial F}{\partial t} - \frac{\partial \psi}{\partial y} \frac{\partial F}{\partial x} + \frac{\partial \psi}{\partial x} \frac{\partial F}{\partial y}. \qquad (3.2)$$

If the scalar field F is materially conserved, then its material derivative must vanish, and $F(x, y, t)$ must satisfy the *convective transport* or *scalar advection* equation:

$$\frac{DF}{Dt} = 0. \qquad (3.3)$$

For example, the scalar Lagrangian label fields a and b trivially satisfy this equation,

$$\frac{Da}{Dt} = \frac{\partial a}{\partial \tau} = 0, \qquad (3.4)$$

as they must, since by definition they are conserved following the flow. Thus, for the steady velocity field $\mathbf{v} = (x, -y)$ of (1.13), with trajectories (1.14) and Lagrangian label fields

$$\mathbf{a}(t; \mathbf{x}) = (x e^{-t}, y e^{t}), \qquad (3.5)$$

the material derivative of the label a is,

$$\frac{Da}{Dt} = \frac{\partial a}{\partial t} + u \frac{\partial a}{\partial x} + v \frac{\partial a}{\partial y} = -x e^{-t} + x e^{-t} + 0 = 0. \qquad (3.6)$$

Equation (3.3) expresses the simple principle that the motion and evolution of a materially conserved property are completely determined by the fluid trajectories. In any real fluid, of course, material conservation of any physical property is not exact, since molecular or turbulent diffusion will also influence property distributions. The focus of the present text is purely on the advective motion, and diffusive effects are therefore neglected. In any physical problem to which these methods are to be applied, the relative importance of advective and diffusive effects must be assessed before the results can be properly interpreted. Even in cases in which diffusion is significant, however, a careful analysis of the advective motion may lead to important insights into the interaction of the advective and diffusive components of the flow. A well-known example of such an interaction is the phenomenon of shear dispersion, which is discussed briefly in the next section.

3.2   Linear Flows

The convective transport equation (3.3) may be solved exactly for the linear motions around stagnation points discussed in Chapter 2. The resulting solutions give insight into the characteristic evolution of materially conserved properties in the different flow regimes of the traveling wave.

When there no flow (case Ia of Chapter 2), it is clear that a materially conserved scalar field F will not change with time.

For the shear flow (case Ib), the equations may be easily solved in transformed (ξ, η) coordinates that are aligned with the flow as in (2.14). In these coordinates, the streamfunction is

$$\psi(\xi, \eta) = -\frac{1}{2}\eta^2, \tag{3.7}$$

and (3.3) takes the form

$$\frac{\partial F}{\partial t} + \eta \frac{\partial F}{\partial \xi} = 0, \tag{3.8}$$

with initial condition

$$F(\xi, \eta, 0) = F_0(\xi, \eta). \tag{3.9}$$

This partial differential equation can be solved by the method of characteristics: the field F must satisfy

$$\frac{d}{ds} F[\xi(s), \eta(s), t(s)] = 0 \tag{3.10}$$

if ξ, η, and t solve the characteristic equations

$$\frac{dt}{ds} = 1, \quad \frac{d\xi}{ds} = \eta, \quad \frac{d\eta}{ds} = 0. \tag{3.11}$$

This yields the solution

$$F(\xi, \eta, t) = F_0(\xi - \eta t, \eta). \tag{3.12}$$

The corresponding scalar gradient is then

$$\frac{\partial F}{\partial \xi} = \frac{\partial F_0}{\partial \xi}(\xi - \eta t, \eta), \tag{3.13}$$

$$\frac{\partial F}{\partial \eta} = \left(\frac{\partial F_0}{\partial \eta} - t\frac{\partial F_0}{\partial \xi} \right)(\xi - \eta t, \eta). \tag{3.14}$$

Thus, the values of the along-flow gradient $\partial F/\partial \xi$ along lines of constant η, say $\eta = \eta_0$, are equal to those of the initial distribution at $\eta = \eta_0$, each translating at constant velocity η_0. The values of the cross-flow gradient $\partial F/\partial \eta$ are composed of the sum of a similarly translating initial cross-flow gradient and a part whose magnitude increases linearly in time at a rate that is proportional to the translating along-flow gradient.

The phenomenon of *shear dispersion* arises from the second term in (3.14): since the shear intensifies the cross-flow gradients, cross-flow diffusion of these gradients by molecular or turbulent processes that are not represented in the pure-shear velocity field will lead to enhanced apparent along-flow diffusion, or dispersion, of the scalar field F. This is a useful example of the interaction between advective and diffusive elements of a flow.

At an elliptic point (case II of Chapter 2; Figure 2.1), the linearized velocity field is a rotation, with streamfunction

$$\psi(\xi, \eta) = \frac{1}{2}\omega(\xi^2 + \eta^2), \tag{3.15}$$

where $\omega > 0$ is a constant. The scalar advection equation can again be solved by the method of characteristics, to yield

$$F(\xi, \eta, t) = F_0(\xi \cos \omega t + \eta \sin \omega t, -\xi \sin \omega t + \eta \cos \omega t). \tag{3.16}$$

Thus, in this linear case, in which the motions along the circular trajectories all have the same period, the scalar gradients rotate, but do not intensify or decay. In the translating frame, the traveling wave has motions of this type around the stagnation points in the channel interior (p_3 and p_4). In the case of a nonlinear rotation around similar circular, or elliptic, trajectories, with shear across adjacent trajectories, the cross-trajectory gradient of F would intensify as in the case of the shear flow above.

At a saddle point (case III of Chapter 2; Figure 2.1), the linearized velocity field is a strain or deformation field, with streamfunction

$$\psi(\xi, \eta) = -\lambda \xi \eta, \tag{3.17}$$

where $\lambda > 0$ is a constant. The scalar advection equation can again be solved by the method of characteristics, to yield

$$F(\xi, \eta, t) = F_0\left(\xi e^{-\lambda t}, \eta e^{\lambda t}\right). \tag{3.18}$$

Thus, $(\partial F/\partial \xi, \partial F/\partial \eta) = (e^{-\lambda t}\partial F_0/\partial \xi, e^{\lambda t}\partial F_0/\partial \eta)$, so the magnitude of the scalar gradient in the diffluent ξ-direction decays exponentially, while that in the confluent η-direction grows exponentially. In the translating frame, the traveling wave has motions of this type near the hyperbolic stagnation points along the channel boundary (p_1, p_2, p_5, and p_6).

The exponential growth of scalar gradients in the strain field near saddle points suggests that these regions and their generalizations in time-dependent flow will be of particular importance for analyses of Lagrangian motions. This proves indeed to be the case, and these generalizations and the associated phenomena form the main subject of this text. Before proceeding, however, it is useful to consider in more detail the Lagrangian motion in the rotational field near elliptic points.

3.3   Closed Material Curves and Integrability

A *closed material curve* is a material curve that forms a simple loop and so may be smoothly deformed to form a circle. Any such curve completely encloses the fluid in its interior. For example, each circular trajectory around the elliptic point of the transformed streamfunction (3.15) is such a closed material curve. The closed material curves that arise in this rotational flow, and their generalizations in time-dependent flow, represent barriers to fluid transport. Their presence indicates that there is no motion of fluid into, or out of, the corresponding enclosed regions.

Trajectories that form nested sets of closed material curves are all essentially equivalent to the circular, periodic trajectories of the linear flow around an elliptic point. In steady, two-dimensional, incompressible flow, it turns out that almost all trajectories—that is, all except some isolated trajectories—consist of closed material curves, provided that certain technical conditions are met. Consider, for example, the unperturbed traveling wave (2.1). It is clear that the trajectories in the recirculation regions (Figure 1.1) are closed material curves. For the traveling wave, however, all of the relevant technical conditions are met only if the periodic channel is viewed in cylindrical geometry, so that the fluid leaving the domain at $x = 2\pi/k$ enters at $x = 0$ (and vice versa). The fluid trajectories in the jet region may then also be regarded as closed material curves. The stagnation points and the material manifolds of the saddle points are not closed material curves, but these represent the allowed, isolated exceptions.

This simple flow structure, in which almost all of the trajectories are nested material curves, is characteristic of *integrable* dynamical systems, of which Lagrangian motion in steady, two-dimensional, incompressible flow is one example. In general, integrable dynamical systems are those for which the equations of motion can be reduced to direct integrations that may be carried out without the need to solve any differential equations. In steady, two-dimensional, incompressible flow, the streamfunction is constant along trajectories,

$$\psi[x(t), y(t)] = \psi[x(0), y(0)] = \psi_0, \tag{3.19}$$

where ψ_0 is a constant. The equation (3.19) implicitly relates $y(t)$ to $x(t)$. If the function $\psi(x, y)$ is known, this relation may in principle be solved to yield

$$y(t) = \mathcal{Y}[x(t)], \tag{3.20}$$

for some function $\mathcal{Y}(x)$. The result may be substituted into the first equation of (1.23) to obtain

$$\frac{dx}{dt} = -\frac{\partial \psi}{\partial y}[x(t), \mathcal{Y}(t)], \tag{3.21}$$

and this may in turn be integrated to yield an implicit equation for $x(t)$,

$$\int^t dt = -\int^x \left[\frac{\partial \psi}{\partial y}[x, \mathcal{Y}(x)]\right]^{-1} dx. \tag{3.22}$$

Upon inversion of this implicit equation, $y(t)$ can be obtained immediately from (3.20). This procedure of solving by substitutions that yield simple integrals is called *reduction to quadratures*. A system that can be completely solved by reduction to quadratures is considered integrable whether or not the integrals can be carried out analytically. In all integrable flows, it can be shown that all except some isolated trajectories form closed material curves, provided again that certain technical conditions are met. Steady, two-dimensional, incompressible flow is always integrable in this sense. Note that the set of equations (3.20)–(3.22) is symbolic: a single functional relation $y = \mathcal{Y}(x)$ can rarely be obtained that is valid for the entire solution, and the inversion must generally be pieced together from several such relations that are valid in different portions of the domain (as in the example in Appendix B).

In contrast, unsteady, incompressible, two-dimensional flow may not be integrable, and the geometry of the fluid trajectories may be much more complex. However, if integrability can be established in that case, even locally, it will have important implications for fluid transport. In such a situation, the trajectories in the integrable region will again form closed material curves, which will represent barriers to fluid transport. Quantifying the conditions under which such curves can be expected to appear in general time-dependent flows would be an important and useful achievement. Unfortunately, this proves to be a difficult task, and only one general result on this problem is available. This result is the Kolmogorov–Arnold–Moser (KAM) theorem. It describes a rigorous, but generally impractical, test for the presence of closed material curves in near-integrable, two-dimensional, incompressible flows. In the following sections, we give a brief summary of this theorem.

3.4 Action-Angle Variables

Consider the transformed streamfunction (3.15). As noted above, the contours of constant ψ are circles. Since the flow is steady, these circles are also streamlines, and so form a nested set of closed material curves that encircle the stagnation point. In polar coordinates (r, θ), where

$$\xi = r \cos \theta, \quad \eta = r \sin \theta, \tag{3.23}$$

the corresponding velocity field is

$$\dot{r} = 0, \quad \dot{\theta} = \omega, \tag{3.24}$$

with solution

$$r = r_0 = \text{constant}, \quad \theta = \omega t + \theta_0 \bmod 2\pi. \tag{3.25}$$

This shows explicitly that all of the solutions near the elliptic point are periodic, with period $T = 2\pi/\omega$.

For the general nonlinear, steady, incompressible, two-dimensional velocity field

$$\dot{x} = -\frac{\partial \psi}{\partial y}(x, y), \quad \dot{y} = \frac{\partial \psi}{\partial x}(x, y), \tag{3.26}$$

a coordinate transformation that gives a similar set of periodic solutions can be sought. Since a smooth coordinate transformation can only distort trajectories, but not change their qualitative character, the existence of such a transformation would demonstrate that a similar structure of nested, closed material curves encircles the stagnation point in the nonlinear flow. This, in turn, would establish that there is no fluid transport into, or out of, the enclosed region surrounding the stagnation point. A coordinate transformation

$$r = R(x, y), \quad \theta = \Theta(x, y) \tag{3.27}$$

with inverse

$$x = \bar{x}(r, \theta), \quad y = \bar{y}(r, \theta) \tag{3.28}$$

would satisfy these conditions if, under this transformation, the equations (3.26) took the form

$$\dot{r} = 0, \quad \dot{\theta} = \Omega(r) \tag{3.29}$$

for some function Ω of r only. The transformed equations (3.29) would then have the solution

$$r = r_0 = \text{constant}, \quad \theta = \Omega(r_0)\, t + \theta_0 \bmod 2\pi. \tag{3.30}$$

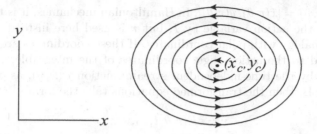

FIGURE 3.1. A region of streamlines in which a transformation to action-angle variables can be constructed. The point (x_c, y_c) is an elliptic stagnation point.

These solutions differ from (3.25) only in that the rate of change of θ may vary from trajectory to trajectory. In both cases the solutions are closed, periodic trajectories, with θ changing linearly with time on each trajectory. The existence of a transformation of this type would be sufficient to demonstrate that the trajectories of the flow (3.26) also form closed material curves in the original (x, y) coordinates. The coordinates (r, θ), in which the transformed equations take the simple form (3.30), are called *action-angle variables*.

A flow, or a local region of a flow, for which the transformation to action-angle variables is possible must have a structure that is qualitatively similar to the linear rotation: it must contain a continuous set of closed periodic trajectories, or contours of constant $\psi(x, y)$ (Figure 3.1). Some additional details of the necessary conditions and the transformation procedure are provided in Appendix B.

3.5   Near-Integrable Fluid Flow

Consider a two dimensional velocity field of the following form:

$$\dot{x} = -\frac{\partial \psi}{\partial y}(x, y) - \varepsilon \frac{\partial \psi_1}{\partial y}(x, y, t, \varepsilon),$$
$$\dot{y} = \frac{\partial \psi}{\partial x}(x, y) + \varepsilon \frac{\partial \psi_1}{\partial x}(x, y, t, \varepsilon),$$

(3.31)

where ε is small, so that $\varepsilon \, \partial \psi_1 / \partial y$ and $\varepsilon \, \partial \psi_1 / \partial x$ are perturbation terms. Since ψ is time-independent, the flow is integrable when $\varepsilon = 0$. For small ε, it is commonly referred to as being *near-integrable*.

For the KAM theorem to apply, the integrable, steady flow must be explicitly expressed in the action-angle coordinate system. In some cases, to which the KAM theorem might otherwise appear relevant, it may not be possible to compute the integrals that define the coordinate transformation (see Appendix B). Assume, however, that it is possible to define a transformation to action-angle variables, $(r, \theta) = [R(x, y), \Theta(x, y)]$, with inverse $(x, y) = [\bar{x}(r, \theta), \bar{y}(r, \theta)]$. In Hamiltonian mechanics, it is traditional to denote the action variable by I, but r is used here instead to emphasize the analogy with polar coordinates. If these coordinates are chosen to correspond to the action-angle coordinates of the integrable, $\varepsilon = 0$ flow described by the time-independent streamfunction $\psi(x, y)$, as outlined in Appendix B, then the transformed equations take the form

$$\dot{r} = \varepsilon F(r, \theta, t, \varepsilon), \tag{3.32}$$
$$\dot{\theta} = \Omega(r) + \varepsilon G(r, \theta, t, \varepsilon),$$

where the explicit time-dependence of the functions F and G is completely determined by the form of ψ_1.

3.6 The KAM Theorem

Consider the transformed flow (3.32). The unperturbed ($\varepsilon = 0$) velocity field has a simple form,

$$\dot{r} = 0,$$
$$\dot{\theta} = \Omega(r), \tag{3.33}$$

with streamfunction

$$\hat{\psi}(r) = \int^r \Omega(r')\, dr' + \text{constant}. \tag{3.34}$$

The trajectories of the unperturbed flow follow circular streamlines $r = r_0 = \text{constant}$, with θ increasing linearly in time at a rate that depends on r_0. Suppose that the perturbation streamfunction ψ_1 in (3.31) depends *quasiperiodically* on time t, with basic frequencies ω_i, $i = 1, \ldots, n$. This is the most general form of perturbation time-dependence for which a KAM theorem has been proven.

The KAM theorem for this quasiperiodically perturbed, transformed flow may be stated as follows. Suppose that for the closed streamline $r = r_0$,

$$\Omega'(r_0) \neq 0. \tag{3.35}$$

This is a nondegeneracy or invertibility condition, which is related to the implicit function theorem of calculus; it means that, near $r = r_0$, the frequency of the periodic motion around an unperturbed closed streamline changes monotonically from streamline to streamline. Suppose also that

$$|m_0 \Omega(r_0) + m_1 \omega_1 + \cdots + m_n \omega_n| \geq \gamma \left(\sum_{i=0}^{n} |m_i| \right)^{-\tau}, \tag{3.36}$$

for all positive or negative nonzero integers m_j, $j = 0, \ldots, n$, and for some $\gamma > 0$ and some $\tau > n$. This is a nonresonance condition, which further constrains the incommensurate frequencies of the quasiperiodic perturbation; in a mathematical sense, however, most sets of basic frequencies ω_i will still satisfy this condition. Suppose also that the perturbation ψ_1 is sufficiently smooth in x and y. Then (KAM theorem), for ε sufficiently small, there exists a closed material curve with the parametric representation

$$r = r_0 + \sqrt{\varepsilon}\chi_1(\theta', \omega_1 t, \ldots, \omega_n t, \sqrt{\varepsilon}),$$
$$\theta = \theta' + \sqrt{\varepsilon}\chi_2(\theta', \omega_1 t, \ldots, \omega_n t, \sqrt{\varepsilon}), \tag{3.37}$$

where the functions $\chi_1(\theta', \omega_1 t, \ldots, \omega_n t, \sqrt{\varepsilon})$ and $\chi_2(\theta', \omega_1 t, \ldots, \omega_n t, \sqrt{\varepsilon})$ are 2π-periodic in θ' and $\omega_i t$, $i = 1, \ldots, n$. Moreover, on this material curve, the trajectories are quasiperiodic with the $n + 1$ basic frequencies $\{\Omega(r_0), \omega_1, \ldots, \omega_n\}$, which (by hypothesis) satisfy the nonresonance condition (3.36).

At each instant of time t, the parametrized curve (3.37) is a closed material curve that traps all the fluid contained in its interior. As time evolves, the shape of this curve may change, as the trajectories oscillate with the frequencies $\{\omega_1, \ldots, \omega_n\}$, but it will always remains a closed material curve, and as such represents a complete barrier to transport. At present, the KAM theorem is the only rigorous mathematical result for proving the existence of such complete barriers to transport in time-dependent flows. From a practical point of view, however, its utility is limited. This is because, for a given flow, the nondegeneracy and nonresonance conditions are difficult to verify, and the rigorous results evidently often underestimate the stability of the flow by a wide margin. Nonetheless, for conceptual reasons, barriers to transport and their dynamical analogues are often referred to as KAM tori, or invariant tori (or KAM or invariant circles, in the two-dimensional case), even when the evidence for them is purely numerical and there is no real possibility of relating them to the KAM theorem itself.

If the nonresonance condition (3.36) is not satisfied for a particular closed streamline $r = r_0$ and perturbation ψ_1, then this streamline is said to be *resonant*. In this case the perturbation will destroy the closed streamline, and fluid trajectories may cross between the corresponding regions in the time-dependent, perturbed flow. The nonresonance condition (3.36) is also referred to as a *Diophantine condition*, and consists of an infinite number of inequalities. Perhaps surprisingly, most frequency vectors are nonetheless Diophantine, and resonance is unusual. The nondegeneracy condition is also satisfied on most streamlines of many geophysical model flows. However, there are interesting situations in which it is violated.

The theorem states that for ε sufficiently small a closed invariant circle exists within a distance on the order of $\sqrt{\varepsilon}$ from the unperturbed closed streamline $r = r_0$, provided the nondegeneracy and nonresonance conditions hold. It does not, however, provide an estimate of the maximum value of ε—that is, the maximum amplitude of the perturbation—for which such a closed invariant circle exists. A number of results have been obtained that give estimates of this maximal perturbation amplitude for the case $n = 1$ above, in which the perturbation has a single basic frequency ω_1.

3.7   Chaos, Integrability, and Advected Scalar Fields

The relevance to geophysical fluid dynamics of the KAM theorem, which was proved originally in the context of classical Hamiltonian mechanics, is part of a broader mathematical analogy between Lagrangian motion in two-dimensional, incompressible, fluid flows, and a particular set of classical Hamiltonian mechanical systems. This analogy is exact: the equations of motion for a mechanical system described by a Hamiltonian function $H(p, q, t)$ can be obtained from the trajectory equations (1.23) by substituting the canonical coordinates p and q and Hamiltonian function H for the spatial coordinates x and y and streamfunction ψ, respectively. Thus, from a mathematical point of view, the study of the fluid trajectories in a given two-dimensional, incompressible fluid flow is equivalent to the study of all possible motions of a given classical time-dependent Hamiltonian system with one degree of freedom.

Any time-dependent, one-degree-of-freedom Hamiltonian system can, in turn, be written as a time-independent, two-degree-of-freedom Hamiltonian system, as follows. Consider the two-degree-of-freedom Hamiltonian system

$$\dot{q}_i = \frac{\partial H}{\partial p_i}, \qquad \dot{p}_i = -\frac{\partial H}{\partial q_i}, \qquad i = 1, 2, \tag{3.38}$$

where

$$H(\mathbf{q}, \mathbf{p}) \equiv \psi(x, y, \tau) - \phi, \tag{3.39}$$

and

$$\mathbf{q} \equiv (q_1, q_2) = (y, \phi), \qquad \mathbf{p} \equiv (p_1, p_2) = (x, \tau). \tag{3.40}$$

Hamilton's equations for this two-degree-of-freedom system are

$$\dot{q}_1 = \dot{y} = \frac{\partial H}{\partial p_1} = \frac{\partial \psi}{\partial x},$$

$$\dot{q}_2 = \dot{\phi} = \frac{\partial H}{\partial p_2} = \frac{\partial \psi}{\partial \tau},$$

$$\dot{p}_1 = \dot{x} = -\frac{\partial H}{\partial q_1} = -\frac{\partial \psi}{\partial y},$$

$$\dot{p}_2 = \dot{\tau} = -\frac{\partial H}{\partial q_2} = 1. \tag{3.41}$$

Hence, (1.23) and (3.38)-(3.40) are equivalent.

An important concept in the study of the motions of classical Hamiltonian systems is that of *chaos*. Strictly speaking, a *chaotic dynamical system* is a dynamical system that contains a chaotic invariant set. A *chaotic invariant set* is a region of phase space—or, in the present context, a portion of the fluid domain—in which there are a countable infinity of periodic trajectories, an uncountable infinity of aperiodic trajectories, and a dense trajectory. A dense trajectory is one that comes arbitrarily close to any other point in the chaotic invariant set. Clearly, chaotic fluid trajectories will be qualitatively different from the integrable trajectories considered above. The combination of dense, aperiodic, and periodic trajectories will lead, for example, to the rapid intensification of gradients of materially conserved scalars, since fluid parcels that are initially separated by finite distances may be brought arbitrarily close together. Like shear dispersion, this advective process is potentially important for the mixing of fluid properties in the presence of diffusion. Thus, it is of interest to know whether chaotic trajectories are possible in a given flow.

In general, it follows from a theorem due to Liouville (e.g., Arnold (1989)), that if a time-independent, two-degree-of-freedom Hamitonian system has two independent integrals of the motion (that is, two materially conserved scalar fields), and if these two functions satisfy certain technical conditions, then the system will be completely integrable. This, with the above equivalence, appears to imply that chaotic fluid motions are impossible in any incompressible, two-dimensional fluid flow with a materially conserved scalar field. However, the technical conditions on the scalar fields involve the definition of the flow domain in both space and time, and may not, in general, be met.

It is straightforward to verify that $H(y, \phi, x, \tau)$ is itself materially conserved,

$$\frac{DH}{Dt} = \frac{\partial H}{\partial x}\dot{x} + \frac{\partial H}{\partial y}\dot{y} + \frac{\partial H}{\partial \tau}\dot{\tau} + \frac{\partial H}{\partial \tau}\dot{\phi}$$
$$= \frac{\partial \psi}{\partial x}\left(-\frac{\partial \psi}{\partial y}\right) + \frac{\partial \psi}{\partial y}\frac{\partial \psi}{\partial x} + \frac{\partial \psi}{\partial \tau} - \frac{\partial \psi}{\partial \tau} = 0. \qquad (3.42)$$

Suppose now that the scalar field $F(x, y, \tau)$ is also materially conserved. The hypotheses of the Liouville theorem require that the two conserved fields F and H must be in *involution*, that is, their Poisson bracket must vanish, where the *Poisson bracket* of F and H is defined by

$$\{F, H\} = \sum_{i=1}^{2}\left(\frac{\partial F}{\partial q_i}\frac{\partial H}{\partial p_i} - \frac{\partial H}{\partial q_i}\frac{\partial F}{\partial p_i}\right). \qquad (3.43)$$

Since

$$\{F, H\} = \frac{\partial F}{\partial y}\frac{\partial H}{\partial x} - \frac{\partial F}{\partial x}\frac{\partial H}{\partial y} + \frac{\partial F}{\partial \phi}\frac{\partial H}{\partial \tau} - \frac{\partial H}{\partial \phi}\frac{\partial F}{\partial \tau}$$

$$= \frac{\partial F}{\partial y}\frac{\partial \psi}{\partial x} - \frac{\partial F}{\partial x}\frac{\partial \psi}{\partial y} + 0 + \frac{\partial F}{\partial \tau}$$

$$= \frac{DF}{Dt}, \tag{3.44}$$

it follows that H and any materially conserved scalar F will be in involution. The functions H and F must satisfy one additional condition: they must be independent, that is, if

$$c_1 \nabla H + c_2 \nabla F = 0, \tag{3.45}$$

where

$$\nabla H = (\psi_x, \psi_\tau, \psi_y, 1), \quad \nabla F = (F_x, F_\tau, F_y, 0), \tag{3.46}$$

then it must follow that

$$c_1 = c_2 = 0. \tag{3.47}$$

Clearly, this condition is always satisfied in any part of the xyt-domain where F is not constant.

This would appear to prove that fluid trajectories for time-dependent, two-dimensional, incompressible flows cannot be chaotic if any materially conserved scalar $F(x, y, t)$ exists. And, since such functions F can always be found (for example, choose F equal to a Lagrangian coordinate a or b in (1.2), as in Section 3.1 above), this in turn appears to imply that trajectories in two-dimensional turbulence cannot be chaotic in a strict sense. In this case, however, appearance is deceptive: neither of these conclusions is true in general. The Liouville theorem depends on an additional technical condition, which demands that in both space and time, the domain on which the flow is defined must be bounded. In a two-dimensional turbulent flow that persists for an infinite time, for example, the time domain is not bounded, and chaotic motion is possible.

In general, chaotic motion is always possible when the domain of the time variable $p_2 = \tau$ cannot be transformed to a bounded region, as can be done in the contrasting case of periodic or quasiperiodic variability. Nonetheless, if the other hypotheses of the Liouville theorem hold, the flow will still be integrable in the sense that it can be reduced to quadratures. This means that, if the materially conserved scalar function $F(x, y, t)$ and the streamfunction $\psi(x, y, t)$ are both known functions of space and time, then the trajectories of the fluid motion can be obtained directly from these two functions in any part of the xyt-domain where F is not constant, without solving the differential equations that determine the trajectories from the velocity field.

3.8 Notes

The original proofs of the KAM theorem were given by Kolmogorov (1954), Arnold (1963), and Moser (1962). The first proofs required that the velocity field be an analytic function. Moser's main contribution was to develop a method of proof for the case that the velocity field has only a finite number of derivatives in the region of interest of the flow. KAM theorems have since been proven for many different types of systems, including three-dimensional incompressible flows and some forms of compressible flows. A useful overview of existing results is the monograph by Broer et al. (1996). The KAM theorem for quasiperiodically time-dependent velocity fields can be found in Kozlov (1996), Arnold et al. (1988), and Jorba & Simo (1996). Note that the velocity field need not be analytic or sufficiently differentiable throughout the entire flow domain; it is enough if the appropriate conditions are satisfied in the local region of interest (for example, in a region of closed streamlines of the unperturbed steady flow). At present there is no KAM theorem for time-dependent velocity fields that are not quasiperiodic.

KAM theory is often cited in connection with regions of periodic or quasiperiodic trajectories observed in numerical simulations or laboratory experiments, but it is rarely possible to demonstrate that the hypotheses of the appropriate theorem are satisfied. In many such cases, the hypotheses may be violated, but the term "KAM torus" is still often used as a qualitative descriptor of regions of periodic or quasiperiodic trajectories embedded in a more complex surrounding Lagrangian flow. Strictly speaking, for example, the unperturbed flow must be transformed into action-angle variables before the nondegeneracy condition (3.35) can be verified, and this is often difficult. In Appendix B, an example is given in which it is possible to carry out this transformation. A more exotic structure that may have some influence on transport in incompressible, two-dimensional, time-periodic flows is a cantorus: a quasiperiodic trajectory that fills out a Cantor set, rather than a closed curve. Cantori and transport have been studied by Aubry (1983b), Aubry (1983a), Mather (1982), and Month & Herrera (1979). There is no theory for cantori in compressible, or three-dimensional flows, or flows in which the explicit time-dependence is not periodic.

Studies of complicated fluid trajectories that arise as a consequence of the violation of the KAM nondegeneracy condition (3.35) have been carried out by del Castillo-Negrete & Morrison (1993), Howard & Humphreys (1995), del Castillo-Negrete et al. (1996, 1997), Voyatzis & Ichtiaroglou (1999), and Petrisor (2001, 2002). Note that this condition is always violated somewhere in jets that have recirculation regions on both sides, such as the traveling-wave example considered here. Results that give estimates of the maximal perturbation amplitude for the case $n = 1$ (i.e., periodic unsteadiness) include Celletti & Chierchia (1988), Herman (1988), de la Llave & Rana (1990), MacKay (1990), MacKay et al. (1989), MacKay &

Percival (1985), Mather (1982, 1984), and Stark (1988). Similarly, it should be clear that KAM theory provides sufficient conditions for the persistence of quasiperiodic trajectories under unsteady perturbations, but does not give complementary conditions for the existence of chaos, or chaotic invariant sets, in nonlinear dynamical systems. Some additional recent work on integrability and materially conserved scalars in geophysical fluid models can be found in Balasuriya (2001).

4

Fluctuating Waves and Meanders

4.1   Time-Dependent Flow in the Comoving Frame

In the translating frame, the undisturbed traveling wave and meandering jet are steady flows. The geometry of the associated fluid trajectories is relatively simple, and has been characterized in detail in the previous chapters. When these flows are disturbed by additional time-dependence, however, the Lagrangian motion associated with the resulting *nonautonomous* velocity fields $\mathbf{v}(\mathbf{x}, t)$ can become extremely complex, even in the translating frame. The development of methods for the analysis of this complex motion is the main subject of this text.

The natural starting point for the examination of Lagrangian motion in time-dependent flow is the generalization of the analysis of steady linearized motion (Sections 2.3 and 2.4) to time-dependent linear motion. Two essential complications arise immediately. First, the linearized motion will generally be governed by a time-dependent velocity field, so the classification of linear motions for steady velocity fields no longer applies. Second, a convenient set of stagnation points often may not be available, so the full geometric analysis will hinge on the identification of another set of distinguished trajectories. The first of these can be addressed by generalizing the corresponding classifications. The second is a more subtle and subjective aspect of the extension of the geometric analysis of Lagrangian motion to time-dependent flows, and is less amenable to systematic, formal analysis; nonetheless, appropriate generalizations are again possible.

For time-dependent linearized velocity fields, the KAM theorem (Chapter 3) provides a conceptual extension of the elliptic-point class of steady linear flows, for which there are nearby trajectories that form closed material curves. In the present chapter, time-dependent generalizations of the saddle-point class of flows are considered. A specific type of distinguished trajectory is identified that plays a fundamental role in the analysis of the Lagrangian motion in the time-dependent traveling-wave flow. Note that, while the KAM theorem applies only for temporally periodic or quasi-periodic near-integrable flows, the time-dependence considered here is not similarly restricted, and the flows need not be nearly steady or integrable. Rather, the concepts and methods developed here apply, in principle, to flows with general time-dependence.

4.2   Linearized Motion: A Time-Dependent Example

For time-dependent linearized velocity fields, it is possible to develop natural generalizations of the saddle-point class of steady linear flows, with its characteristic pair of exponentially growing and decaying solutions. Even in the linear case, however, the time-dependent velocity field must in general be considered directly in order to obtain qualitatively correct characterizations of the corresponding trajectories. These two basic aspects of Lagrangian motion in time-dependent linearized motion can be illustrated with a simple example.

Let the time-dependent velocity field be given by $\mathbf{v}(\mathbf{x}, t) = M(t)\mathbf{x}$, where

$$M(t) = \begin{pmatrix} \cos 4t & \sin 4t - 2 \\ \sin 4t + 2 & -\cos 4t \end{pmatrix}, \tag{4.1}$$

and consider the trajectories given by

$$\dot{\mathbf{x}} = M(t)\mathbf{x}. \tag{4.2}$$

Clearly, the origin $(x, y) = (0, 0)$ is a stagnation point, and so also a trajectory. The velocity field $M(t)\mathbf{x}$ is the superposition of a steady pure rotation $A\mathbf{x}$ and a time-dependent pure strain $B(t)\mathbf{x}$, where

$$A = \begin{pmatrix} 0 & -2 \\ 2 & 0 \end{pmatrix}, \qquad B(t) = \begin{pmatrix} \cos 4t & \sin 4t \\ \sin 4t & -\cos 4t \end{pmatrix}. \tag{4.3}$$

The rotation matrix A has eigenvalues $\pm 2i$. The strain matrix $B(t)$ has time-independent eigenvalues ± 1, but time-dependent eigenvectors, corresponding to a deformation field of constant strength but changing orientation.

Since the velocity field is linear, the matrix associated with its linearization is just the matrix $M(t) = A + B(t)$ itself. This matrix also has time-independent eigenvalues, and its real Jordan canonical form is

$$M_C = \begin{pmatrix} 0 & -\sqrt{3} \\ \sqrt{3} & 0 \end{pmatrix}, \tag{4.4}$$

which would appear to generate purely rotational trajectories. Naively, one might be tempted to conclude that the stagnation point at the origin must an elliptic point, hence stable, and that fluid elements initially near the origin would always remain near the origin. Alternatively, one might hope to determine the character of the trajectories for long times $(t \to \infty)$ by considering the time average of M, which is simply A. This again is a pure rotation, and again one might be tempted to conclude that the stagnation point must be stable and that nearby fluid elements will remain nearby.

Both of these conclusions would be false. The first argument fails because the corresponding eigenvectors are time-dependent, so solutions of $\dot{\mathbf{x}} = M_C \mathbf{x}$ do not correspond to solutions of (4.2). The second argument fails because, for a given solution $\mathbf{x}(t)$, (4.2) is nonlinear in t despite being linear in $\mathbf{x}$. Consequently, the time-averaged trajectories of the time-dependent velocity field may differ from the trajectories of the time-averaged velocity field.

The correct trajectories for (4.2) can be obtained by introducing a time-dependent coordinate rotation,

$$\begin{pmatrix} \xi \\ \eta \end{pmatrix} = \begin{pmatrix} \cos 2t & \sin 2t \\ -\sin 2t & \cos 2t \end{pmatrix} \begin{pmatrix} x \\ y \end{pmatrix}. \tag{4.5}$$

In these rotating coordinates, the equations for the trajectories become

$$\dot{\xi} = \xi, \quad \dot{\eta} = -\eta, \tag{4.6}$$

which are easily solved. The transformation (4.5) may then be inverted to obtain the trajectories

$$x(t) = a e^t \cos 2t - b e^{-t} \sin 2t,$$
$$y(t) = a e^t \sin 2t + b e^{-t} \cos 2t. \tag{4.7}$$

Hence trajectories with $a = x(0) = 0$ and $b = y(0) \neq 0$ approach the origin exponentially, but trajectories with $a \neq 0$ eventually depart from it exponentially. Thus, the stagnation point has the character of a saddle point, not an elliptic point.

4.3 Linearization About a Trajectory

Consider now the linearization of a time-dependent velocity field about an arbitrary trajectory, rather than about a stagnation point as in Chapter 2. Let the time-dependent trajectory be given by $\mathbf{x} = \mathbf{x}_0(t)$, and let a nearby trajectory be denoted by $\mathbf{x} = \tilde{\mathbf{x}}(t) = \mathbf{x}_0(t) + \mathbf{x}'(t)$. The linear approximation may be obtained essentially as before, by substituting this expression for $\tilde{\mathbf{x}}(t)$ in the equation $\dot{\mathbf{x}} = \mathbf{v}(\mathbf{x}, t)$, expanding in a Taylor series in $\mathbf{x}'$, and neglecting terms of quadratic and higher order in $\mathbf{x}'$. Note that, since $\mathbf{x}_0(t)$ is a trajectory, the terms $\dot{\mathbf{x}}_0(t) = \mathbf{v}[\mathbf{x}_0(t), t]$ will cancel after the Taylor expansion. The result is

$$\dot{\mathbf{x}}' = D\mathbf{v}[\mathbf{x}_0(t), t] \, \mathbf{x}', \tag{4.8}$$

or, equivalently, in matrix form,

$$\begin{pmatrix} \dot{x}' \\ \dot{y}' \end{pmatrix} = \begin{pmatrix} \dfrac{\partial u}{\partial x}[\mathbf{x}_0(t)] & \dfrac{\partial u}{\partial y}[\mathbf{x}_0(t)] \\[2ex] \dfrac{\partial v}{\partial x}[\mathbf{x}_0(t)] & \dfrac{\partial v}{\partial y}[\mathbf{x}_0(t)] \end{pmatrix} \begin{pmatrix} x' \\ y' \end{pmatrix}, \tag{4.9}$$

where the partial derivatives are evaluated at $\mathbf{x} = \mathbf{x}_0(t)$, and the matrix $D\mathbf{v}(\mathbf{x}_0(t), t)$ is again the *velocity gradient tensor* (or the *Jacobian matrix* or *matrix associated with the linearization*) about $\mathbf{x}_0(t)$. As before, the determinant of this matrix vanishes for incompressible flow. Note that time-dependence of the linearized velocity field $D\mathbf{v}(\mathbf{x}_0(t), t)\mathbf{x}'$ can arise in two different ways: through the explicit dependence of $\mathbf{v}$ on t, but also through the dependence of $D\mathbf{v}$ on the time-dependent solution $\mathbf{x}_0(t)$. Thus, the linearized velocity field about a trajectory may be time-dependent even when the original velocity field is steady. Consequently, time-dependent generalizations are required even for the linear analysis of motion near general trajectories of steady flows.

In general, the nonautonomous linear differential equations (4.9) have two linearly independent solutions $\phi_j = \mathbf{x}_j(t)$, $j = 1, 2$ (e.g., Coddington & Levinson (1955)). The general solution of (4.9) can be written in terms of the *fundamental solution matrix* $\Phi(t) = (\phi_1^T, \phi_2^T)$, the columns of which are these two linearly independent solutions:

$$\Phi(t) = \begin{pmatrix} x_1(t) & x_2(t) \\ y_1(t) & y_2(t) \end{pmatrix}. \tag{4.10}$$

Of course, this is true also for autonomous equations. For example, the fundamental solution matrices for cases II and III of Chapter 2 may be written

$$\Phi(t) = \Phi_{\text{II}}(t) = \begin{pmatrix} \cos \omega t & -\sin \omega t \\ \sin \omega t & \cos \omega t \end{pmatrix} \tag{4.11}$$

and

$$\Phi(t) = \Phi_{\text{III}}(t) = \begin{pmatrix} e^{\lambda t} & 0 \\ 0 & e^{-\lambda t} \end{pmatrix}, \tag{4.12}$$

respectively. Since in these two examples $\Phi(0)$ is simply the identity matrix, the corresponding solutions for arbitrary initial conditions (ξ_0, η_0) are then given by

$$\begin{pmatrix} \xi(t) \\ \eta(t) \end{pmatrix} = \Phi(t) \begin{pmatrix} \xi_0 \\ \eta_0 \end{pmatrix}. \tag{4.13}$$

In general, a *principal fundamental solution matrix* (Chicone (1999)) $X(t, t_1)$ can be constructed with a similar property. Namely, with the definition

$$X(t, t_1) = \Phi(t)\Phi^{-1}(t_1), \tag{4.14}$$

where t_1 is an arbitrary, fixed initial time and $\Phi^{-1}(t)$ is the inverse of $\Phi(t)$, $X(t, t_1)$ is again a fundamental solution matrix, and $X(t_1, t_1)$ is always equal to the identity matrix. The general solution with

$$\xi(t_1) = \xi_1, \quad \eta(t_1) = \eta_1, \tag{4.15}$$

may then be written

$$\begin{pmatrix} \xi(t) \\ \eta(t) \end{pmatrix} = X(t, t_1) \begin{pmatrix} \xi_1 \\ \eta_1 \end{pmatrix}. \tag{4.16}$$

This linearized trajectory is the linear approximation to the trajectory passing through the point $\mathbf{x} = \mathbf{x}_0(t_1) + (\xi_1, \eta_1)$ at $t = t_1$. For incompressible flow, it follows from the considerations in Section 1.2 that the determinant $\det J = \det X(t, t_1)$ is equal to 1 for all t and t_1.

4.4 Lyapunov Exponents

The example in Section 4.2 illustrated that the qualitative behavior of trajectories in a time-dependent linearized flow cannot be inferred from the time average of the velocity field. However, a useful quantity, the *Lyapunov exponent*, can be computed from an average of the linearized trajectories themselves.

Given the principal fundamental solution matrix $X(t, t_1)$ for the linearized equations (4.8) about a trajectory $\mathbf{x}_0(t)$, with $\mathbf{x}_0(t_1) = \mathbf{x}_1$, consider the linearized trajectory $X(t, t_1)\mathbf{e}_1$ with initial condition $\mathbf{e}_1 = (\xi_1, \eta_1)$ at $t = t_1$. Then, the corresponding *Lyapunov exponent* $\lambda(\mathbf{x}_0, \mathbf{e}_1, t_1)$ is the limit

$$\lambda(\mathbf{x}_0, \mathbf{e}_1, t_1) = \lim_{t \to \infty} \frac{1}{t - t_1} \ln \|X(t, t_1)\mathbf{e}_1\|. \tag{4.17}$$

Here $\|\mathbf{x}\| := \sqrt{\mathbf{x} \cdot \mathbf{x}}$ denotes the magnitude of $\mathbf{x}$, so the Lyapunov exponent measures the average exponential rate of growth of the distance from $\mathbf{x}_0(t)$ to $\mathbf{x}_0(t) + X(t, t_1)\mathbf{e}_1$, the linearized approximation to the trajectory passing through $\mathbf{x}_1 + \mathbf{e}_1$ at $t = t_1$.

The Lyapunov exponent λ depends on the trajectory $\mathbf{x}_0(t)$ and the direction of the initial vector $\mathbf{e}_1$ at $t = t_1$ on the linearized trajectory about $\mathbf{x}_0(t)$. However, it does not depend on the amplitude of $\mathbf{e}_1$, since, for an arbitrary constant C,

$$\begin{aligned}
\lambda(\mathbf{x}_0, C\mathbf{e}_1, t_1) &= \lim_{t \to \infty} \frac{1}{t - t_1} \ln C\|X(t, t_1)\mathbf{e}_1\| \\
&= \lim_{t \to \infty} \frac{1}{t - t_1} (\ln C + \ln \|X(t, t_1)\mathbf{e}_1\|) \\
&= \lambda(\mathbf{x}_0, \mathbf{e}_1, t_1). \tag{4.18}
\end{aligned}$$

Also, for a given linearized trajectory about $\mathbf{x}_0(t)$, the value of λ does not depend on which initial point is chosen. That is, suppose $\mathbf{e}_2 = X(t_2, t_1)\mathbf{e}_1$, so that $(\xi_2, \eta_2) = \mathbf{e}_2$ is a point on the same linearized trajectory as the point $(\xi_1, \eta_1) = \mathbf{e}_1$. Then, with (4.14), which shows that $X(t, t_2)X(t_2, t_1) = X(t, t_1)$,

$$\begin{aligned}
\lambda(\mathbf{x}_0, \mathbf{e}_1, t_1) &= \lim_{t \to \infty} \frac{1}{t - t_1} \ln \|X(t, t_2)X(t_2, t_1)\mathbf{e}_1\| \\
&= \lim_{t \to \infty} \left(\frac{1}{t - t_2} - \frac{t_2 - t_1}{(t - t_1)(t - t_2)} \right) \ln \|X(t, t_2)\mathbf{e}_2\| \\
&= \lambda(\mathbf{x}_0, \mathbf{e}_2, t_2). \tag{4.19}
\end{aligned}$$

Thus, λ depends only on the linearized trajectory identified by the initial point, not on the location of that initial point along the given trajectory.

Now, suppose e_1 and e_2 are two different initial vectors at $t = t_1$, such that $\lambda_1 = \lambda(x_0, e_1, t_1) > \lambda_2 = \lambda(x_0, e_2, t_1)$. Then, from (4.17),

$$\lim_{t \to \infty} \frac{\|X(t, t_1)e_2\|}{\|X(t, t_1)e_1\|} = \lim_{t \to \infty} e^{-(\lambda_1 - \lambda_2)t} = 0. \tag{4.20}$$

It follows that, for all constants C_1 and C_2 not both zero, $\lambda(x_0, C_1 e_1 + C_2 e_2, t_1) = \lambda_1$ if $C_1 \neq 0$, and $\lambda(x_0, C_1 e_1 + C_2 e_2, t_1) = \lambda_2$ only if $C_1 = 0$. Thus, for a given trajectory of a given linearized, two-dimensional flow, λ may attain at most two different values. For incompressible flow, for which $\det X(t_1, t_2) = 1$ for all t_1 and t_2, those two values are either both zero, or a pair of equal and opposite positive and negative numbers. In the latter case, for which then $\lambda_2 = -\lambda_1 < 0$, a perhaps surprising consequence is that the definition of the Lyapunov exponents identifies a unique exponentially decaying direction, but not a unique exponentially growing direction: the decaying direction e_s is determined by $e_s = e_2$, since $C_1 = 0$ is necessary for decay ($\lambda = -\lambda_1$), while any solution $e_1 + C_2 e_2$, with C_2 arbitrary, will grow ($\lambda = \lambda_1$).

Consider now two examples. First, the Lyapunov exponents of the stagnation point trajectory $x_0 = (0, 0)$ in the time-dependent flow (4.2) can be found directly from the explicit solution (4.7): the Lyapunov exponent in the direction $(0, 1)$ at time $t = 0$ is -1, and that in any other direction at time $t = 0$ is $+1$. Similarly, the Lyapunov exponents of a saddle-type stagnation point in steady flow will be the eigenvalues of the linearized velocity field.

Second, consider a region around an elliptic point, where the velocity field can be expressed in action-angle variables,

$$\dot{r} = 0, \quad \dot{\theta} = \Omega(r), \tag{4.21}$$

and the trajectories are given by

$$r = r_0 = \text{constant}, \tag{4.22}$$
$$\theta(t) = \Omega(r_0)\, t + \theta_0.$$

Linearizing (4.21) about (4.22) gives

$$\begin{pmatrix} \dot{\xi} \\ \dot{\eta} \end{pmatrix} = \begin{pmatrix} 0 & 0 \\ \Omega'(r_0) & 0 \end{pmatrix} \begin{pmatrix} \xi \\ \eta \end{pmatrix}. \tag{4.23}$$

The principal fundamental solution matrix of (4.23) is

$$X(t, t_1) = \begin{pmatrix} 1 & 0 \\ \Omega'(r_0)(t - t_1) & 1 \end{pmatrix}. \tag{4.24}$$

Let $\mathbf{e}_1 = (0, 1)$ and $\mathbf{e}_2 = (1, 0)$. Then

$$\lambda[(r_0, \theta_0), \mathbf{e}_1] = 0,$$
$$\lambda[(r_0, \theta_0), \mathbf{e}_2] = 0,$$

for any (r_0, θ_0) labeling a trajectory of (4.21) defined by (4.22). Thus, the Lyapunov exponents of the periodic trajectories in any integrable region of a two-dimensional flow will be exactly zero.

4.5 Exponential Dichotomies

The definition of the Lyapunov exponent is a useful step toward the desired time-dependent generalization of the saddle-point classification. It introduces the limit $t \to \infty$ to replace the steady-flow eigenvalue analysis, and it uniquely identifies one set of asymptotically exponential solutions. That set describes the exponentially decaying direction $\mathbf{e}_s$ of the linearized time-dependent flow; this is $\mathbf{e}_2$ in the example (4.20). The set of linearized solutions with initial conditions $C_2 \mathbf{e}_2$, for any constant C_2, corresponds to the set of linearized trajectories that are asymptotic to the original trajectory $\mathbf{x}_0(t)$ as $t \to \infty$, and is thus analogous to the exponentially decaying solution from the eigenvalue analysis (2.14) of the linear steady flow.

This calculation, however, does not uniquely identify an exponentially growing counterpart to the decaying direction, which would be analogous to the exponentially growing solution from the eigenvalue analysis of the linear steady flow: in the example (4.20), all directions $\mathbf{e}_1 + C_2 \mathbf{e}_s$ grow exponentially, for any C_2. To obtain this counterpart set of trajectories, it is necessary to consider, in addition, the equivalent of the Lyapunov exponent in the opposite temporal limit $t \to -\infty$. This allows the identification of the solutions that decay toward $\mathbf{x}_0(t)$ in that opposite limit, and these turn out to be an adequate counterpart set of trajectories. The consideration of the two limits $t \to \pm\infty$ and the identification of the two corresponding sets of trajectories can be combined into a single construction, the *exponential dichotomy*. The approach is illustrated here first, followed by the formal definition and two examples.

Consider the time-dependent linear flow (4.2). In matrix form, the solutions (4.7) are

$$\mathbf{x}(t) = Y(t)\mathbf{a}, \quad Y(t) = \begin{pmatrix} e^t \cos 2t & -e^{-t} \sin 2t \\ e^t \sin 2t & e^{-t} \cos 2t \end{pmatrix}, \tag{4.25}$$

where $\mathbf{a} = (a, b)$, and $Y(t)$ is the fundamental solution matrix for (4.2). Note that $Y(0) = $ id, the identity matrix.

Clearly, one or the other of the two linearly independent solutions may be isolated in (4.25) by setting either a or b to zero. Formally, this can be done by defining a matrix *projection* Q, where

$$Q = \begin{pmatrix} 0 & 0 \\ 0 & 1 \end{pmatrix}. \tag{4.26}$$

Consider the two orthogonal operators Q and $(\mathrm{id} - Q)$. These will take an arbitrary initial vector $\mathbf{a}$ to the corresponding initial condition for one of the two linearly independent solutions:

$$Q\mathbf{a} = \begin{pmatrix} 0 \\ b \end{pmatrix}, \qquad (\mathrm{id} - Q)\mathbf{a} = \begin{pmatrix} a \\ 0 \end{pmatrix}. \tag{4.27}$$

Because of the time-dependence of the solutions, however, these operators cannot be applied directly to $\mathbf{x}(t) = Y(t)\mathbf{a}$ at an arbitrary time t, with the same effect. Instead, the solution must first be inverted to obtain the corresponding initial condition, and the projection operator then applied to this initial condition:

$$QY^{-1}(t)\mathbf{x}(t) = Q\mathbf{a} = \begin{pmatrix} 0 \\ b \end{pmatrix},$$

$$(\mathrm{id} - Q)Y^{-1}(t)\mathbf{x}(t) = (\mathrm{id} - Q)\mathbf{a} = \begin{pmatrix} a \\ 0 \end{pmatrix}. \tag{4.28}$$

The two linearly independent solutions at time t can then be reconstructed by applying the time-evolution operator $Y(t)$:

$$Y(t)QY^{-1}(t)\mathbf{x}(t) = be^{-t} \begin{pmatrix} -\sin 2t \\ \cos 2t \end{pmatrix},$$

$$Y(t)(\mathrm{id} - Q)Y^{-1}(t)\mathbf{x}(t) = ae^{t} \begin{pmatrix} \cos 2t \\ \sin 2t \end{pmatrix}. \tag{4.29}$$

In this way, the fundamental solution matrix $Y(t)$ and the projection operator Q can be combined to separate a general trajectory $\mathbf{x}(t) = Y(t)\mathbf{a}$ into its linearly independent components. As before, it is clear that the stagnation point $(0, 0)$ in this example has saddle-type stability.

The definition of an exponential dichotomy involves a similar, slightly generalized construction. Given the principal fundamental solution matrix $X(t)$ (or, in general, $X(t, t_1)$, but here for simplicity it is assumed that $t_1 = 0$ and the dependence on t_1 is dropped from the notation), an *exponential dichotomy* exists if a projection operator P and suitable constants K_1, K_2, α, $\beta > 0$ can be found such that

$$\| X(t)PX^{-1}(\tau) \| \leq K_1 \exp\left(-\alpha(t - \tau)\right), \quad t \geq \tau, \tag{4.30}$$

$$\| X(t)(\mathrm{id} - P)X^{-1}(\tau) \| \leq K_2 \exp\left(\beta(t - \tau)\right), \quad t \leq \tau. \tag{4.31}$$

Here $\| \cdot \|$ denotes a matrix norm, such as the maximum of the absolute values of the matrix elements. For two-dimensional, incompressible flow, the projection operators P and $\mathrm{id} - P$ must each identify one and only one linearly independent solution; in general, the sum of their ranks must equal the dimension of the full system (4.2). Note that the projection procedure used in (4.30) and (4.31) has been generalized slightly from the above example by the introduction of the second time variable τ, to remove the explicit dependence on the distinguished initial time $t = 0$.

As a simple example of a velocity field possessing an exponential dichotomy, consider the linear, steady, two-dimensional velocity field around the hyperbolic point p_1 in the unperturbed traveling wave:

$$\dot{x} = -\lambda x,$$
$$\dot{y} = \lambda y, \qquad (4.32)$$

where $\lambda > 0$. The fundamental solution matrix is given by

$$X(t) = \begin{pmatrix} e^{-\lambda t} & 0 \\ 0 & e^{\lambda t} \end{pmatrix}, \qquad (4.33)$$

and an appropriate choice of P is

$$P = \begin{pmatrix} 1 & 0 \\ 0 & 0 \end{pmatrix}.$$

Then

$$X(t)PX^{-1}(\tau) = \begin{pmatrix} e^{-\lambda(t-\tau)} & 0 \\ 0 & 0 \end{pmatrix},$$

and

$$X(t)\left(\mathrm{id} - P\right)X^{-1}(\tau) = \begin{pmatrix} 0 & 0 \\ 0 & e^{\lambda(t-\tau)} \end{pmatrix}.$$

So the choices $K_1 = K_2 = 1$, $\alpha = \beta = \lambda$ satisfy (4.30) and (4.31).

As a second example, consider (4.2). As shown above, the solution $\mathbf{x} = (0,0)$ has the character of a saddle point, so this system should also have an exponential dichotomy. Let $X = Y$ and $P = Q$, where Y and Q are as defined above in (4.25) and (4.26), so that

$$X(t)PX^{-1}(\tau) = e^{-(t-\tau)} \begin{pmatrix} \sin 2\tau \sin 2t & -\cos 2\tau \sin 2t \\ -\sin 2\tau \cos 2t & \cos 2\tau \cos 2t \end{pmatrix}, \qquad (4.34)$$

from which it follows that

$$\| X(t)PX^{-1}(\tau) \| \le e^{-(t-\tau)}, \qquad t \ge \tau. \qquad (4.35)$$

Similarly,

$$X(t)\left(\mathrm{id} - P\right)X^{-1}(\tau) = e^{(t-\tau)} \begin{pmatrix} \cos 2\tau \cos 2t & \sin 2\tau \cos 2t \\ \cos 2\tau \sin 2t & \sin 2\tau \sin 2t \end{pmatrix}, \qquad (4.36)$$

from which it follows that

$$\| X(t) \, (\text{id} - P) \, X^{-1}(\tau) \| \leq e^{(t-\tau)}, \qquad t \leq \tau. \tag{4.37}$$

Taking $K_1 = K_2 = 1$ and $\alpha = \beta = 1$ shows that (4.2), as anticipated, has an exponential dichotomy. Note that (4.2) is an example of a system of (linear) ordinary differential equations with time-periodic coefficients but no periodic solutions (except, of course, the trivial solution $\mathbf{x} = (0,0)$).

4.6   Hyperbolic Trajectories

Stagnation points of saddle type were shown in Chapter 2 to have a special significance for Lagrangian motion in steady, incompressible, two-dimensional flow. Material manifolds associated with these points, and with the exponentially growing and decaying trajectories of the corresponding linearized velocity fields, form boundaries between neighboring regions of qualitatively different Lagrangian motion.

To extend the geometric analysis of Lagrangian motion to time-dependent velocity fields, it is necessary to identify analogues of these saddle-type stagnation points and their associated material manifolds. A natural and appropriate analogue of the hyperbolic stagnation point is a *hyperbolic trajectory*. (Recall that that saddle points are the incompressible, two-dimensional form of the more general mathematical class of hyperbolic stagnation points.) A hyperbolic trajectory is a trajectory $\gamma(t) = \mathbf{x}_0(t)$ for which the linearized velocity field $D\mathbf{v}[\mathbf{x}_0(t), t]$ about $\mathbf{x}_0(t)$ has an exponential dichotomy. Thus, in each of the two examples in the preceding section, the stagnation point $\mathbf{x} = (0,0)$ is a simple example of a hyperbolic trajectory. The exponential dichotomy guarantees the existence of two sets of asymptotically exponential trajectories of the linearized velocity field, one approaching $\mathbf{x}_0(t)$ as $t \to \infty$ and the other approaching $\mathbf{x}_0(t)$ as $t \to -\infty$, and provides a projection P that identifies these two trajectories. These sets of asymptotically exponential trajectories form the material manifolds that will provide the foundation of the subsequent geometric analysis of Lagrangian motion in time-dependent flows.

Note that it is the properties of the surrounding trajectories, obtained from the linearized velocity field, that make a given trajectory hyperbolic. These properties can be described and interpreted geometrically. Consider again the example (4.32). The exponential dichotomy was demonstrated by decomposing the trajectories into components associated with the two fundamental solutions in (4.33), one exponentially decaying and the other exponentially growing. The set of exponentially decaying solutions is just the x axis, and the set of exponentially growing solutions is the y axis. Physically, this means that, for the flow (4.32) the x axis and y axis are each material curves; mathematically, they are one-dimensional linear subspaces of the plane, determined by the respective fundamental solutions, which are invariant under the flow. These subspaces are denoted by E^s and E^u, respectively: the stable subspace E^s corresponds to the set of decaying solutions (the x axis, in this example) and the unstable subspace E^u corresponds to the set of growing solutions (the y axis).

A similar decomposition is possible for general hyperbolic trajectories. In this case, the sets of decaying and growing trajectories will again be material curves, which at each fixed time t are again straight lines and so form one-dimensional linear subspaces $E^s(t)$ and $E^u(t)$ of the plane. However, the orientations of these lines in general now depend on the time t. In this case, the fundamental solution matrix $X(t)$ and the projection operator P can be combined with the sets of all possible initial conditions to obtain explicit expressions for the linear subspaces:

$$E^s(t) = \{\text{all } \mathbf{e}_s(t) \text{ such that } \mathbf{e}_s(t) = X(t)P\mathbf{a} \text{ for some } \mathbf{a}\}, \tag{4.38}$$

$$E^u(t) = \{\text{all } \mathbf{e}_u(t) \text{ such that } \mathbf{e}_u(t) = X(t)(\text{id} - P)\mathbf{a} \text{ for some } \mathbf{a}\}. \tag{4.39}$$

Here $\mathbf{e}_s(t)$ and $\mathbf{e}_u(t)$ are the vectors in the respective subspaces at time t, and $\mathbf{a}$ runs over all possible initial vectors. Note that since $X(t)$ is invertible, $E^s(t)$ and $E^u(t)$ remain distinct for all t if they are so at any t; that is, the vectors $\mathbf{e}_s(t)$ and $\mathbf{e}_u(t)$ cannot become collinear.

Inequalities describing the asymptotic decay and growth of these sets of trajectories can then be obtained directly from (4.30) and (4.31). Namely, since

$$\mathbf{e}_s(t) = X(t)PX^{-1}(\tau)\mathbf{e}_s(\tau) \tag{4.40}$$

and

$$\mathbf{e}_u(t) = X(t)\left(\text{id} - P\right)X^{-1}(\tau)\mathbf{e}_u(\tau), \tag{4.41}$$

it follows that

$$\|\mathbf{e}_s(t)\| \leq K_1 e^{-\alpha(t-\tau)}\|\mathbf{e}_s(\tau)\|, \;\; t \geq \tau, \tag{4.42}$$

$$\|\mathbf{e}_u(t)\| \leq K_2 e^{\beta(t-\tau)}\|\mathbf{e}_u(\tau)\|, \;\; t \leq \tau, \tag{4.43}$$

for some constants $K_1, K_2, \alpha, \beta > 0$. These inequalities show that any trajectory through $E^s(\tau)$ asymptotically approaches the origin at an exponential rate as $t \to \infty$, and any trajectory through $E^u(\tau)$ asymptotically approaches the origin at an exponential rate as $t \to -\infty$. These are the linearized representations of trajectories that asymptotically approach $\mathbf{x}_0(t)$ as $t \to \infty$ and $t \to -\infty$, respectively.

Thus, at each time t the linearized field of motion around a hyperbolic trajectory is divided into four regions by a pair of material lines, $E^s(t)$ and $E^u(t)$, whose orientations change with time and that intersect at the hyperbolic trajectory (Figure 4.1). These material lines consist of the sets of points through which pass the sets of trajectories that approach the hyperbolic trajectory $\mathbf{x}_0(t)$ asymptotically as $t \to \infty$ and $t \to -\infty$, respectively. Since they are material lines, other fluid trajectories cannot cross them, and they form natural, time-dependent boundaries of flow regimes.

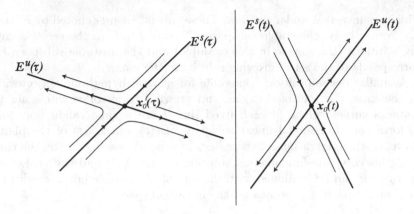

FIGURE 4.1. Geometry of the stable and unstable subspaces of the linearized system associated with the hyperbolic trajectory $\gamma(t) = \mathbf{x}_0(t)$. The left panel shows the stable and unstable subspaces at time $t = \tau$, and the right panel shows them at some later time $t > \tau$. Note that the subspaces may translate and rotate in space as time evolves.

Consider the time-dependent linear flow (4.2) above. The stagnation point at the origin has an exponential dichotomy, so $\gamma(t) = \mathbf{x}_0(t) = (0,0)$ is a simple example of a hyperbolic trajectory. Let $\mathbf{a} = (a, b)$ range over all possible initial conditions at $t = 0$ for a linearized trajectory about $\mathbf{x}_0(t)$. At time t, the stable subspace $E^s(t)$ is given by

$$E^s(t) = \{\mathbf{e}_s(t) = X(t)P\mathbf{a}\} = \left\{ a \begin{pmatrix} -\sin 2t \\ \cos 2t \end{pmatrix} : -\infty < a < \infty \right\}. \quad (4.44)$$

Similarly, at time t, the unstable subspace $E^u(t)$ is given by

$$E^u(t) = \{\mathbf{e}_u(t) = X(t)(\mathrm{id} - P)\mathbf{a}\} = \left\{ b \begin{pmatrix} \cos 2t \\ \sin 2t \end{pmatrix} : -\infty < b < \infty \right\}. \quad (4.45)$$

For fixed t, $E^s(t)$ and $E^u(t)$ are thus each one-dimensional subspaces through $\gamma(t) = (0,0)$, the orientations of which depend on t.

4.7   Material Manifolds of Hyperbolic Trajectories

In the case of hyperbolic stagnation points in steady flow, it was shown in Chapter 2 that there are material manifolds of the nonlinear flow $\mathbf{v}(\mathbf{x})$ that have the same asymptotic properties as the stable and unstable subspaces E^s and E^u of the linearized flow $D\mathbf{v}(\mathbf{x}_0)$, and are tangent to these subspaces at the stagnation point $\mathbf{x}_0$. A similar set of material manifolds can be defined for hyperbolic trajectories in time-dependent nonlinear flow. In this case, the corresponding time-dependent material manifolds $W^s(t)$ and $W^u(t)$ are material curves that are tangent at each time t to the time-dependent linear subspaces $E^s(t)$ and $E^u(t)$, respectively, of the linearized flow $D\mathbf{v}[\mathbf{x}_0(t)]$ at the hyperbolic trajectory $\gamma(t) = \mathbf{x}_0(t)$.

In general, the stable and unstable manifolds $W^s(t)$ and $W^u(t)$ of a hyperbolic trajectory $\mathbf{x}_0(t)$, and the associated linear subspaces $E^s(t)$ and $E^u(t)$, all depend both on time t and on the trajectory $\mathbf{x}_0(t)$. It can be useful to indicate just one of these dependencies explicitly in the notation, and suppress the other, or to indicate both explicitly. Thus, the different notations $W^s(t)$, $W^s[\mathbf{x}_0(t)]$, $W^s[\mathbf{x}_0]$, and $W^s[t, \mathbf{x}_0(t)]$ can all denote the same time-dependent and trajectory-dependent material manifold, but emphasize different aspects of these dependencies, as appropriate in each context. The same is true of the corresponding notations for W^u, E^s, and E_u.

The *stable material manifold* $W^s[\mathbf{x}_0(t)]$ consists of all the points through which pass, at time t, the trajectories of the nonlinear velocity field $\mathbf{v}(\mathbf{x}, t)$ that are asymptotic to the hyperbolic trajectory $\mathbf{x}_0(t)$ as $t \to \infty$. The *unstable material manifold* $W^u[\mathbf{x}_0(t)]$ consists of all the points through which pass, at time t, the trajectories of the nonlinear velocity field $\mathbf{v}(\mathbf{x}, t)$ that are asymptotic to the hyperbolic trajectory $\mathbf{x}_0(t)$ as $t \to -\infty$. In the case of steady flow, the existence of these material manifolds for the nonlinear velocity field was a consequence of the Morse lemma. Here, it is established by the *stable and unstable manifold theorem for hyperbolic trajectories*.

Hyperbolic trajectories $\mathbf{x}_0(t)$ and their stable and unstable manifolds $W^s[\mathbf{x}_0(t)]$ and $W^u[\mathbf{x}_0(t)]$ are robust structures, in a precise mathematical sense: if they exist for a given velocity field, and that field is then altered, they will always exist for the altered field, provided that the alteration is sufficiently small. These properties are established by the *persistence*

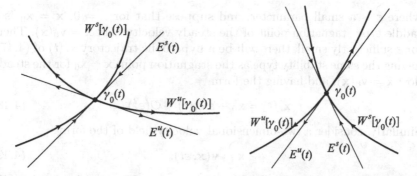

FIGURE 4.2. Geometry of the stable and unstable manifolds of the nonlinear system associated with the hyperbolic trajectory $\gamma(t) = \mathbf{x}_0(t)$. The left panel shows the stable and unstable manifolds at time τ, and the right panel shows them at some later time $t > \tau$. Note that the manifolds may translate and rotate in space as time evolves.

theorem for hyperbolic trajectories and stable and unstable manifolds, which may be summarized as follows. Consider the velocity field

$$\dot{\mathbf{x}} = \mathbf{v}(\mathbf{x}, t; \varepsilon), \tag{4.46}$$

where ε is a parameter or vector of parameters, and

$$\mathbf{v}(\mathbf{x}, t; 0) = \mathbf{v}(\mathbf{x}, t).$$

Suppose that at $\varepsilon = 0$, (4.46) has a hyperbolic trajectory $\mathbf{x}_0(t)$ with stable and unstable material manifolds $W^s[\mathbf{x}_0(t)]$ and $W^u[\mathbf{x}_0(t)]$. Then, for sufficiently small ε, (4.46) will have a hyperbolic trajectory $\mathbf{x}_\varepsilon(t)$ with stable and unstable material manifolds $W^s[\mathbf{x}_\varepsilon(t)]$ and $W^u[\mathbf{x}_\varepsilon(t)]$, which will have the same properties as $W^s[\mathbf{x}_0(t)]$ and $W^u[\mathbf{x}_0(t)]$, with $\mathbf{x}_0(t)$ replaced by $\mathbf{x}_\varepsilon(t)$.

4.8   Hyperbolic Trajectories for Near-Integrable Flows

The definition of hyperbolic trajectories and their associated material manifolds provides an adequate extension of the saddle-point classification of steady flows to flows with general time-dependence, from which the subsequent geometric analysis of Lagrangian motion can proceed. The next step is to develop methods to identify the specific distinguished hyperbolic trajectories around which to conduct the analysis.

In the case of near-integrable flow, the persistence properties of hyperbolic trajectories often provide useful guidance. For example, consider a two-dimensional velocity field of the form

$$\dot{\mathbf{x}} = \mathbf{v}_0(\mathbf{x}) + \varepsilon \mathbf{v}_1(\mathbf{x}, t), \tag{4.47}$$

where ε is a small parameter, and suppose that for $\varepsilon = 0$, $\mathbf{x} = \mathbf{x}_0$ is a saddle-type stagnation point of the steady velocity field $\dot{\mathbf{x}} = \mathbf{v}_0(\mathbf{x})$. Then, for ε sufficiently small, there will be a hyperbolic trajectory $\mathbf{x}_\varepsilon(t)$ of (4.47), having the same stability type as the stagnation point $\mathbf{x} = \mathbf{x}_0$ for the steady flow $\dot{\mathbf{x}} = \mathbf{v}_0(\mathbf{x})$, and having the form

$$\mathbf{x}_\varepsilon(t) = \mathbf{x}_0 + \varepsilon \mathbf{x}_1(t) + \mathcal{O}(\varepsilon^2). \tag{4.48}$$

Similarly, consider a two-dimensional velocity field of the form

$$\dot{\mathbf{x}} = \mathbf{v}(\mathbf{x}, \varepsilon t), \tag{4.49}$$

where again ε is considered a small parameter. This can be rewritten in the equivalent three-dimensional form

$$\dot{\mathbf{x}} = \mathbf{v}(\mathbf{x}, z),$$
$$\dot{z} = \varepsilon. \tag{4.50}$$

Suppose that for $\varepsilon = 0$ and for all z, (4.50) has a continuous set of saddle-type stagnation points of the form

$$\mathbf{x} = \mathbf{x}_0(z).$$

Then, for ε sufficiently small, there will be a hyperbolic trajectory $\mathbf{x}_\varepsilon(t)$ of (4.47), having the same stability type as the set of stagnation points $\mathbf{x} = \mathbf{x}_0(z)$ of $\dot{\mathbf{x}} = \mathbf{v}(\mathbf{x}, z)$, and time-derivative $|\dot{\mathbf{x}}_\varepsilon(t)|$ of order ε. Morever, the trajectory $\mathbf{x} = \mathbf{x}_\varepsilon(t)$ will lie within an $\mathcal{O}(\varepsilon)$ distance of $\mathbf{x} = \mathbf{x}_0(z)$ for each z.

4.9   The Traveling Wave

Consider now the time-dependent traveling wave flow (1.30). In the reference frame moving with the primary wave, the streamfunction for the time-dependent traveling wave takes the form

$$\psi(x, y, t) = \psi_0(x, y) + \varepsilon\psi_1(x, y, t)$$
$$= -cy + A\sin kx\ \sin y + \varepsilon\psi_1(x, y, t). \qquad (4.51)$$

The associated velocity field in the comoving frame

$$\dot{x} = c - A\sin kx\ \cos y - \varepsilon\frac{\partial\psi_1}{\partial y}(x, y, t), \qquad (4.52)$$

$$\dot{y} = Ak\cos kx\ \sin y + \varepsilon\frac{\partial\psi_1}{\partial x}(x, y, t)., \qquad (4.53)$$

is in the near-integrable form (4.47). Thus, hyperbolic trajectories can be identified near the saddle points of the integrable ($\varepsilon = 0$) flow as outlined in Section 4.8.

As a specific example, suppose that the time-dependent disturbance streamfunction ψ_1 has the form

$$\psi_1(x, y, t) = y\, e^{-\delta|t|}. \qquad (4.54)$$

Physically, this corresponds to a flow in which a pulse of uniform along-channel flow is superposed on the traveling wave, with maximum perturbation at $t = 0$ and a temporal pulse width of $2/\delta$. This example is the simplest of several that will be explored in more detail in the succeeding chapters.

Recall that, for the traveling wave flow with $\varepsilon = 0$, there are four saddle-type stagnation points p_1, p_2, p_5, and p_6, on the boundaries $y = \{0, \pi\}$ of the channel. For small ε, there will be a time-dependent hyperbolic trajectory $\mathbf{x}_\varepsilon(t) = [x_\varepsilon(t), y_\varepsilon(t)]$ near each of these four points. Since $\partial\psi_1/\partial x = 0$, each of these trajectories will have $y_\varepsilon(t) = y_0 = \{0, \pi\}$, with motion only in x, along the boundary. Substituting this and the form

$$x_\varepsilon(t) = x_0 + \varepsilon x_1(t) + \mathcal{O}(\varepsilon^2), \qquad (4.55)$$

into (4.52), where $x_0 = x_s/k$ for the saddle point p_1, expanding, and equating terms of equal order in ε yields

$$\dot{x}_1 = -\alpha x_1 - e^{-\delta|t|}, \tag{4.56}$$

where $\alpha = Ak\delta_s = k(A^2 - c^2)^{1/2}$. The approximation to the relevant time-dependent hyperbolic trajectory $\mathbf{x}_\varepsilon(t)$ is obtained by seeking the corresponding solution for x_1 that remains bounded for all time ($-\infty < t < +\infty$), since this will satisfy the criterion that the hyperbolic trajectory $\mathbf{x}_\varepsilon(t)$ remain within an order-ε distance of p_1. The result is

$$x_1(t) = \begin{cases} -(\alpha + \delta)^{-1}e^{\delta t}, & t \le 0, \\ -(\alpha + \delta)^{-1}e^{-\alpha t} + (\alpha - \delta)^{-1}(e^{-\alpha t} - e^{-\delta t}), & t > 0, \end{cases} \tag{4.57}$$

with the resulting solution

$$\mathbf{x}_\varepsilon(t) = [x_0 + \varepsilon x_1(t) + \mathcal{O}(\varepsilon^2), 0] \tag{4.58}$$

for the hyperbolic trajectory. The trajectory $\mathbf{x}_\varepsilon(t)$ thus consists of a drift along the channel boundary from x_0 toward negative x during $-\infty < t < 0$, followed by a return drift toward x_0 during $0 < t < \infty$. The maximum excursion of $\mathbf{x}_\varepsilon(t)$ from $\mathbf{x}_0$ is a distance $\varepsilon(\alpha + \delta)^{-1} + \mathcal{O}(\varepsilon^2)$ toward negative x, which is reached (to first order in ε) at $t = 0$.

The corresponding stable and unstable subspaces $E^s[\mathbf{x}_\varepsilon(t)]$ and $E^u[\mathbf{x}_\varepsilon(t)]$, which form the local linear approximations to the time-dependent stable and unstable material manifolds $W^s[\mathbf{x}_\varepsilon(t)]$ and $W^u[\mathbf{x}_\varepsilon(t)]$ of the hyperbolic trajectory, respectively, can be obtained by linearizing (4.52) and (4.53) around $\mathbf{x}_\varepsilon(t)$. To first order in ε, this linearization yields the same result as (2.18). Thus, $E^s[\mathbf{x}_\varepsilon(t)]$ and $E^u[\mathbf{x}_\varepsilon(t)]$ are described by the vectors $\mathbf{e}_s = a(1, 0)$ for all a and $\mathbf{e}_u = b(0, 1)$ for all b, respectively. Note that, although these vectors define constant directions for $E^s[\mathbf{x}_\varepsilon(t)]$ and $E^u[\mathbf{x}_\varepsilon(t)]$, the entire subspaces themselves move in x with the hyperbolic trajectory $\mathbf{x}_\varepsilon(t)$, since they consist of trajectories of the form $\mathbf{x}(t) = \mathbf{x}_\varepsilon(t) + \mathbf{x}'(t)$, where $\mathbf{x}'(t)$ is the small departure from $\mathbf{x}_\varepsilon(t)$ described by the linearized solutions.

The hyperbolic trajectories and stable and unstable subspaces associated with the other three saddle points of the traveling wave flow can be computed in the same way as above for p_1. Expressions for the corresponding material manifolds $W^s(t)$ and $W^u(t)$ of the nonlinear flow are not easily obtained in analytical form. However, as outlined in the next chapter, adequate numerical approximations to $W^s(t)$ and $W^u(t)$ can be computed using linearized estimates of initial conditions from the expressions for $E^s(t)$ and $E^u(t)$, which for each t are tangent at $\mathbf{x}_\varepsilon(t)$ to $W^s(t)$ and $W^u(t)$, respectively.

4.10 Notes

Mathematical aspects of linearization about a specific trajectory are discussed in Arnold (1973), Hirsch & Smale (1974), and Wiggins (2003). Relations between Lyapunov exponents and exponential dichotomies are discussed in Dieci et al. (1997) and Ide et al. (2002). Fundamental references for Lyapunov exponents and exponential dichotomies are Oseledec (1968) and Coppel (1978), respectively; for the latter, see also Henry (1981) and Massera & Schäffer (1966). An excellent elementary but careful introduction to Lyapunov exponents is given by Legras & Vautard (1996). Some recent attempts have been made to generalize Lyapunov exponents to include nonlinear effects (see, e.g., Boffetta et al. (2001)); currently, these generalizations are purely computational and without rigorous mathematical foundation. An interesting application of this approach to transport in the Mediterranean Sea is described in d'Ovidio et al. (2004).

Many numerical methods have been developed for accurately computing Lyapunov exponents; see, e.g., Dieci et al. (1997), Dieci & Eirola (1999), Bridges & Reich (2001), Greene & Kim (1987), and Geist et al. (1990). Issues relating to the convergence of numerical estimates to the asymptotic Lyapunov exponent are discussed in Goldhirsch et al. (1987), Dieci et al. (1997), and Lapeyre (2002). Related work can be found in Janaki et al. (1999) and Rangarajan et al. (1998). These numerical approaches involve either a QR or a singular value decomposition (SVD) of the fundamental solution matrix. The fundamental solution matrix varies in time, which requires extensions of the well-known theory of QR and SVD for constant matrices. These extensions are developed in Dieci & Eirola (1999) and Dieci & Vleck (1999). Methods for computing Lyapunov exponents that are specific to Hamiltonian systems can be found in Partovi (1999) and Yamaguchi & Iwai (2001). Ramasubramanian & Sriram (2000) compare several different algorithms for the computation of Lyapunov exponents.

The *stable and unstable manifold theorem* for hyperbolic stagnation points and hyperbolic periodic trajectories is well known, and statements of the theorem can be found in many textbooks. The stable and unstable manifold theorem for hyperbolic trajectories for velocity fields having arbitrary time-dependence is less well known, but can be obtained from simple modifications of results found in Coddington & Levinson (1955) and Hale (1980). The theorem in this form can be found in Kaper (1992); see also Yi (1993). A discrete-time version can be found in Katok & Hasselblatt (1995). Less well known works that give similar results are Irwin (1973) and de Blasi & Schinas (1973). Recent numerical and theoretical approaches to the definition and identification of hyperbolic trajectories and their stable and unstable manifolds in finite-time velocity fields include Haller (2000), Haller (2001), Haller & Yuan (2000), Haller (2002), Ide et al. (2002), Mancho et al. (2003), Mancho et al. (2004), and Ju et al. (2003).

The persistence of hyperbolic structures under perturbation allows their approximation by regular perturbation theory for small amplitude time-dependent perturbations to steady flow. This approach was followed in this chapter for the traveling wave example. For slowly varying time-dependence, the problem is more delicate, and an approach based on singular perturbation theory is required. Related work in the area of control theory can be found in Desoer (1969) and in Peuteman et al. (2000); the latter contains useful historical references for control applications. In the context of fluid mechanics (Stokes flows) such results were used in Kaper (1992). Recently, these results were further extended in Haller & Poje (1998). In the purely mathematical context related results can be found in Coppel (1978) (pp. 50 and 54).

5

Material Manifolds, Flow Regimes, and Fluid Exchange

5.1 Fluid Exchange and Lobes

In this chapter and the next, it is shown how transport and exchange between flow regimes in time-dependent flows can be understood and quantified by analyzing the structure of the material manifolds of hyperbolic trajectories. With the previous chapters as preparation, a detailed consideration of the phenomena sketched in Figures 1.1 and 1.2 for the meandering jet is possible for the traveling wave. The discussion is organized around a set of similar, but quantitative, figures for the traveling wave, based on the numerical calculations of Malhotra & Wiggins (1998).

Consider first the following conceptual example (Figure 5.1). Suppose that for all $t < 0$, the velocity field is the undisturbed traveling-wave flow in the comoving frame (2.1), but that at time $t = 0$, there is an additional instantaneous, finite displacement of the fluid along the channel that interrupts the undisturbed flow. Suppose further that the traveling-wave velocity field is not itself displaced by this interruption, and that it resumes in a continuous manner immediately after $t = 0$ (so that the $t \to 0^-$ and $t \to 0^+$ limits of the velocity field are equal), and that the velocity field in the comoving frame continues to be equal to the undisturbed traveling-wave flow (2.1) for all $t > 0$.

During each of the intervals $-\infty < t < 0$ and $0 < t < \infty$, the manifolds and streamlines of the undisturbed traveling-wave flow (Figure 1.27) will describe the Lagrangian motion in this example. Thus, just before $t = 0$, the fluid enclosed in the recirculation regime that is defined for (2.1) by the stable and unstable material manifolds of the two hyperbolic points p_1 and p_2 will have been confined there for the entire interval $-\infty < t < 0$. None of this trapped fluid will have crossed into the jet regime (Figure 5.1a). At $t = 0$, however, the instantaneous finite displacement that interrupts the traveling-wave flow will abruptly shift some of this fluid out of the recirculation and into the jet regime, and simultaneously shift an equal amount from the jet into the recirculation regime, along its opposite side (Figure 5.1b). When the steady traveling-wave flow resumes immediately after $t = 0$, the resulting patterns of fluid motion can again be described by these manifolds and streamlines of the undisturbed flow, and the patches of

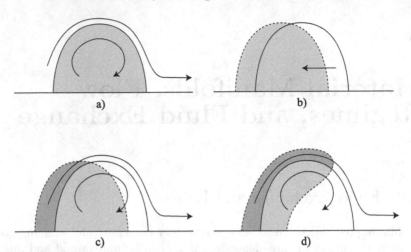

FIGURE 5.1. Conceptual example of fluid exchange induced in the traveling wave flow by an impulsive translation at $t = 0$. (a) For $-\infty < t < 0$, the steady flow streamlines describe the flow, and the fluid in the recirculation (light shading) is trapped. (b) At $t = 0$, all of the fluid is impulsively displaced upstream. (c) For $t > 0$, the steady flow streamlines again describe the flow. (d) Thus, for $0 < t < \infty$, a patch of previously trapped fluid (dark shading) is swept downstream in the jet, while another patch of fluid (unshaded region within closed steady streamline) previously in the jet remains trapped in the recirculation, along with the remainder (light shading) of the originally trapped fluid.

shifted fluid will be confined to their new regimes for all $t > 0$ (Figure 5.1c). The fluid that has been moved from the recirculation region into the jet region will continue downstream toward increasing x as $t \rightarrow \infty$, while the fluid that has been moved from the jet region into the recirculation region will remain trapped in the recirculation as $t \rightarrow \infty$. In this way, the time-dependent perturbation—in this case, the instantaneous, finite displacement at $t = 0$—to the steady flow causes an exchange of fluid between regimes that cannot occur in the steady flow itself.

This conceptual example illustrates two fundamental points. First, time-dependence can induce fluid exchange that is not possible in steady flow. Second, that exchange involves the transport of fluid from one flow regime to another, not from one fixed region of space to another. In this example, in which the flow in the comoving frame is steady except at the single instant $t = 0$, the regimes are also fixed in space in the comoving frame. In the original frame of reference, however, neither regime corresponds to a fixed spatial region. The essential distinction between the recirculation the jet regimes is the differing qualitative pattern of motion in each regime. In this simple example, in which the velocity field is steady for all t except at the instant $t = 0$, the fixed material manifolds of the hyperbolic stagnation points of the undisturbed steady flow form useful delimiters of the two flow regimes, even when the flow is disturbed by the instantaneous displacement at $t = 0$. In general, however, a description intrinsic to the time-dependent flow is necessary. It turns out that such a description can be developed using the material manifolds of hyperbolic trajectories of the time-dependent flow. The corresponding patches of fluid, whose borders consist of specially chosen segments of the material manifolds, will be called *lobes*. The following sections introduce this point of view and the associated mathematical concepts and definitions, using the example of the traveling-wave flow to develop and illustrate the approach.

The analysis of fluid exchange accomplished by the development and evolution of these lobes has been called *lobe dynamics* (Rom-Kedar et al. (1990), Rom-Kedar & Wiggins (1990), Wiggins (1992), Beigie et al. (1994), Malhotra & Wiggins (1998)). We use the term *lobe transport* here instead, to avoid confusion with the dynamics of physical balances according to which geophysical fluid motions evolve. Our purpose in this chapter and the next is to give an accessible introduction to the quantitative methods of lobe transport. These methods and ideas lend themselves naturally to geometrical description. In order to follow this material, it is important to have a firm grasp of the precise conceptual content of the elements of illustrations such as Figures 1.1 and 1.2. The reader is encouraged to think carefully about the meaning of these illustrations and those in the numerical examples to follow as this discussion proceeds.

5.2   Transient Disturbances

Consider the time-dependent traveling-wave flow (1.29), with a simple continuous analogue of the abrupt displacement in the conceptual example above (Figure 5.1). In (1.29), the time-dependent component in the co-moving frame is the disturbance streamfunction ψ_1. For ψ_1, consider the choice

$$\psi_1(x, y, t) = y\, e^{-\delta|t|}. \tag{5.1}$$

This corresponds to a time-dependent velocity field, superposed on the traveling wave, with uniform flow along the channel toward negative x and with amplitude $\varepsilon e^{-\delta|t|}$ at each fixed t. Instead of the instantaneous finite displacement in the conceptual example, this pulse of along-channel flow is distributed over a time-interval around $t = 0$ with an exponential half-width of $1/\delta$.

The hyperbolic trajectory $\mathbf{x}_\varepsilon(t)$ near the undisturbed saddle stagnation point p_1 was obtained in Section 4.9, along with expressions for the corresponding stable and unstable subspaces $E^s[\mathbf{x}_\varepsilon(t)]$ and $E^u[\mathbf{x}_\varepsilon(t)]$. Similar trajectories and subspaces can be obtained near the other three saddle points, p_2, p_5, and p_6. Denote these four hyperbolic trajectories by $\gamma_j(t)$, $j = 1, \ldots, 4$, where γ_j is the trajectory near p_j for each j, so that $\gamma_1 = \mathbf{x}_\varepsilon$. For each γ_j, nonlinear approximations to the corresponding stable and unstable material manifolds $W^s[\gamma_j] = W^s[\gamma_j(t)]$ and $W^u[\gamma_j] = W^u[\gamma_j(t)]$ can be computed from the linear approximations $E^s[\gamma_j(t)]$ and $E^u[\gamma_j(t)]$, using, for example, the numerical methods discussed in Appendix C. At time $t = 0$, $W^s[\gamma_2]$ and $W^u[\gamma_1]$ form symmetric looping structures that divide the plane into regions (Figure 5.2). These regions qualitatively resemble the patches sketched for the conceptual example of instantaneous displacement (Figure 5.1). However, the regions are now defined in terms of the material manifolds of the time-dependent flow, without direct reference to the original, unperturbed, steady traveling-wave flow. This allows the following quantitative identification of the patches as *lobes*. This definition is intrinsic to the time-dependent flow and can be generalized and extended to other such flows in a straightforward manner.

First, recall that, at a fixed time t, the material manifolds $W^s(t)$ and $W^u(t)$ are composed of continuous sets of points that move with the fluid. In a time-dependent flow, their structure at a fixed time t does not, for example, itself represent a fluid trajectory. Thus, two such manifolds $W^s(t)$ and $W^u(t)$ may intersect. In contrast, two different fluid trajectories may not intersect at a given time t, since such an intersection would mean that two different fluid elements with different histories and different futures would occupy the same point at the same time, violating both a fundamental principle of classical physics and the property of uniqueness of solutions to the corresponding mathematical problem. Points of intersection of the stable and unstable manifold pair $W^s[\gamma_2]$ and $W^u[\gamma_1]$ have a special meaning: since each such manifold is defined by the asymptotic properties of trajectories that pass through it, such points of intersection must have the asymptotic properties of both manifolds. This means that a trajectory passing through one of these points must both asymptotically approach the hyperbolic trajectory γ_2 as $t \to \infty$ and asymptotically approach the hyperbolic trajectory γ_1 as $t \to -\infty$. This distinguishes these points from all others on each of these manifolds, for which the corresponding trajectories will have only one of these two asymptotic properties.

Note that, in the steady ($\varepsilon = 0$) traveling-wave flow (Figure 1.3), the situation is different: the stable manifold $W^s(p_2)$ and the unstable manifold $W^u(p_1)$ coincide, forming a single contiguous boundary between the recirculation region and the jet. In this steady case, all of the points in these manifolds are intersection points, and the boundary is a fixed material curve, which may also be interpreted as a trajectory, since each trajectory on each of these two manifolds traces out the same path in space (though not in time). This special structure in the steady flow is degenerate, in the sense that almost any time-dependent perturbation to the flow will destroy the coincidence of the manifolds, and change the qualitative structure so that any intersection points of the two manifolds are isolated, if they exist at all. Note that as $\varepsilon \to 0$, $\gamma_1(t) \to p_1$, $\gamma_2(t) \to p_2$, $W^s[\gamma_2(t)] \to W^s(p_2)$, and $W^u[\gamma_1(t)] \to W^u(p_1)$, as discussed in Sections 4.8 and 4.9. Thus, it is

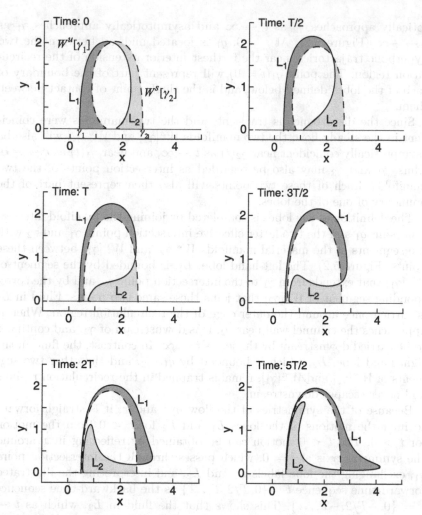

FIGURE 5.2. Material (unstable, solid line; stable, dashed) manifolds for the time-dependent traveling-wave flow with $\psi_1 = y e^{-\delta|t|}$, $\delta = 1.0$, and $\varepsilon = 0.3$. The manifolds are shown at times that are integer multiples of $T/2$, where $T = 2$. The lobes L_1 (dark shading) and L_2 (light shading) are indicated. Only finite segments of the manifolds are shown, contiguous with the corresponding hyperbolic trajectories.

appropriate to discuss the different manifold structures for $\varepsilon = 0$ and $\varepsilon > 0$ as changes in form of given, ε-dependent manifolds.

For time-dependent traveling-wave flow, with the particular choice (5.1) of time-dependent disturbance ψ_1, it turns out that, at any fixed time t, there is exactly one intersection point $q_1(t)$ of the two material manifolds $W^s[\gamma_2]$ and $W^u[\gamma_1]$, corresponding to exactly one trajectory that asymptotically approaches γ_2 as $t \to \infty$ and asymptotically approaches γ_1 as $t \to -\infty$ (Figure 5.2). At $t = 0$, q_1 is located midway between the two hyperbolic trajectories, near the farthest interior extension of the recirculation region. The point $q_1(t = 0)$ will represent part of the boundary of each of the lobes defined below, and is the single point of contact between them.

Since the disturbance is transient, and the two manifolds were coincident in the steady flow, the two manifolds $W^s[\gamma_2]$ and $W^u[\gamma_1]$ will also be asymptotically coincident near $\gamma_2(t)$ as $t \to \infty$, and near $\gamma_1(t)$ as $t \to -\infty$. Thus, γ_1 and γ_2 may also be regarded as intersection points of the two manifolds. Each of these two points will also then represent part of the boundary of one of the lobes.

The definition of the lobes is completed by joining the manifold intersection point q_1 and the two latter effective intersection points γ_1 and γ_2 with the segments of the material manifolds $W^s[\gamma_2]$ and $W^u[\gamma_1]$ between these points (Figure 5.2). The left-hand lobe, L_1, is bounded by the segment of $W^u[\gamma_1]$ that extends from γ_1 to the intersection point q_1, and by the corresponding segment of $W^s[\gamma_2]$ that joins these same two points. Fluid in L_1 is carried once around the outer edge of the recirculation region. When it approaches the channel wall near γ_2, it is downstream of γ_2, and continues to be carried downstream by the jet as $t \to \infty$. In contrast, the fluid in the right-hand lobe, L_2, which is bounded by q_1, γ_2, and the other two segments of $W^s[\gamma_2]$ and $W^u[\gamma_1]$, remains trapped in the recirculation regime, and never escapes downstream.

Because of the symmetries of the flow in t and x, it is straightforward to infer the motions of the lobes L_1 and L_2 for $t < 0$ from the motion for $t > 0$. The $t < 0$ motion can be obtained by reflecting in x around the symmetry axis at $t = 0$, which passes through the intersection point q_1, exchanging the lobe labels L_1 and L_2, and interpreting the illustrated forward-time sequence $t = \{0, T/2, T, \dots\}$ as the backward-time sequence $t = \{0, -T/2, -T, \dots\}$. This shows that the fluid in L_2, which as $t \to \infty$ remains trapped in the recirculation, originates far upstream toward negative x, outside the recirculation region, while the fluid in L_1, which escapes in the jet as $t \to \infty$, originates from the recirculation region.

Thus, the effect of the time-dependent disturbance ψ_1 (5.1) is to entrain a lobe of upstream fluid (L_2) from the upstream jet into the trapped recirculation, and to eject a lobe of trapped fluid (L_1) from the recirculation into the downstream jet. This is precisely the behavior anticipated in the conceptual example (Figure 5.1), in which the time-dependent disturbance was concentrated in a finite, instantaneous displacement of the fluid at $t = 0$. For ψ_1 as in (5.1) and Figure 5.2, the areas of the lobes L_1 and L_2 can be computed directly from the numerical solutions, providing a quantitative measure of the amplitude of exchange between the jet and recirculation that is induced by the time-dependent disturbance.

In this example, the distinction between the jet and the recirculation

region is intuitively clear. The reader will note, however, that these regions
have not yet been defined with reference to the time-dependent material manifolds. Such a definition can be achieved as follows. As $t \to \infty$,
$W^s[\gamma_2(t)]$ increasingly resembles the boundary between the jet and recirculation in the steady ($\varepsilon = 0$) flow. Thus, $W^s[\gamma_2(t)]$ provides a natural definition of the regime boundary for $t > 0$. On the other hand, as $t \to -\infty$,
$W^u[\gamma_1(t)]$ increasingly resembles the steady-flow regime boundary. Thus,
$W^u[\gamma_1(t)]$ provides a natural definition of the regime boundary for $t < 0$.
A complete time-dependent regime boundary $B(t)$ can be constructed by
combining these two:

$$B(t) = \begin{cases} W^u[\gamma_1(t)], & t \leq 0, \\ W^s[\gamma_2(t)], & t > 0. \end{cases} \tag{5.2}$$

Note that, during either interval $-\infty < t < 0$ or $0 < t < \infty$, when the
boundary $B(t)$ is defined by a single material manifold, no exchange of
fluid across the boundary is possible, because trajectories cannot cross material curves. Denote by $R(t)$ the region enclosed by $B(t)$ and the channel
wall at $y = 0$. With respect to this definition of the region $R(t)$ and its
boundary $B(t)$, all of the fluid exchange occurs instantaneously at $t = 0$,
when the regime boundary is redefined. At $t = 0$, the lobe L_1 moves from
the interior of R bounded by $B(t)$, where it had been enclosed by $W^u[\gamma_1(t)]$,
to the exterior of R, where it is confined by $W^s[\gamma_2(t)]$. Similarly, the lobe
L_2 moves at $t = 0$ from the exterior to the interior of $R(t)$. In the lobe-
transport approach to the analysis of fluid exchange, in which lobes and
fluid regimes are defined by boundaries constructed from material manifolds of hyperbolic trajectories, it is always the case that fluid exchange can
occur only when the time-dependent regime boundary is redefined. Note
that, when the time-dependent flow is time-periodic, as in Section 5.3 below, rather than transient, as here, this basic fact can be obscured by a
natural choice of the redefinition of the time-dependent boundary, which,
surprisingly, gives the appearance that the boundary is fixed in time.

For the simple transient disturbance (5.1), the material manifolds
$W^u[\gamma_1(t)]$ and $W^s[\gamma_2(t)]$ intersect, at a fixed time t, at only the single isolated point $q_1(t)$. When more complicated time-dependence is introduced

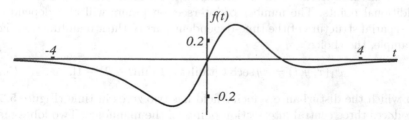

FIGURE 5.3. Temporal structure $f(t) = \mathrm{sech}\, t\, [\tanh t - \tanh(t - 2) - 1]$ of the
disturbance (5.3).

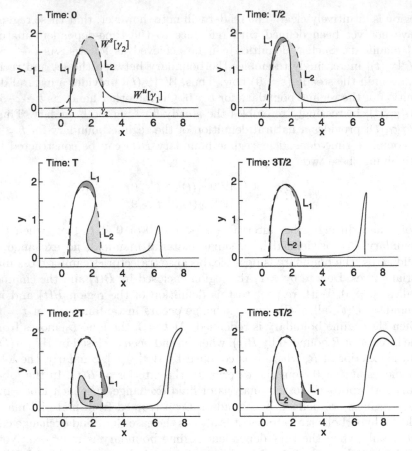

FIGURE 5.4. Material (unstable, solid line; stable, dashed) manifolds for the time-dependent traveling-wave flow with $\psi_1 = xy\,\mathrm{sech}\,t\,[\tanh t - \tanh(t-2) - 1]$ and $\varepsilon = 0.3$. The manifolds are shown at times that are integer multiples of $T/2$, where $T = 2$. The lobes L_1 (dark shading) and L_2 (light shading) are indicated. Only finite segments of the manifolds are shown, contiguous with the corresponding hyperbolic trajectories.

in the disturbance flow, the resulting material manifolds will intersect at additional points. The number of intersection points will also depend on the spatial structure of the time-dependent part of the streamfunction. For example, the choice

$$\psi_1(x, y, t) = xy\,\mathrm{sech}\,t\,[\tanh t - \tanh(t-2) - 1], \tag{5.3}$$

for which the disturbance velocity field has two zeros in time (Figure 5.3), produces three central intersection points of the manifolds. Two lobes may then be defined that are bounded by the pairs of segments of manifolds that join each neighboring pair of these three intersection points (Figure 5.4).

Again, fluid in the left-hand lobe at $t = 0$ is swept downstream, while that in the right-hand lobe remains confined to the recirculation regime. If a time-dependent regime boundary is constructed from the material manifolds as in the example (5.1) above, the lobe transport will again be seen to be associated with the time-dependent redefinition of this boundary. The number of lobes that may be defined will increase with each increase in the number of intersection points of the manifolds, which is often in turn related to the number of oscillations in the time-dependent disturbance, if this disturbance takes a sufficiently simple form. Generally, the lobes of interest are of the latter type described above in this example: they are bounded by pairs of segments of the manifolds that join neighboring pairs of intersection points, and are not adjacent to the hyperbolic trajectories themselves, as they were in the first example.

It may be helpful to repeat here the admonition given at the end of the preceding section: The reader is encouraged to review carefully these two examples and the meaning of Figures 5.2 and 5.4. A firm grasp of the concepts illustrated in these examples should make the details of the subsequent sections easier to follow.

5.3 Oscillatory Disturbances

When the disturbance flow is oscillatory, the number of manifold intersection points may become infinite. In this case, an infinite number of lobes may be defined. However, despite this increased complexity, a lobe analysis can still be carried out that provides a quantitative description of fluid exchange, and simultaneously provides a systematic method for defining the regimes between which fluid is exchanged in each specific case.

In the traveling-wave flow (1.29), consider a disturbance field ψ_1 that is itself a traveling wave with periodic time-dependence,

$$\psi_1(x, y, t) = \sin k_1(x - c_1't)\sin l_1 y. \tag{5.4}$$

Here $c_1' = (c - c_1)$ is the phase speed in the comoving frame, k_1 and l_1 are x- and y-wavenumbers, respectively, and $T = 2\pi/(k_1 c_1')$ is the oscillation period in the comoving frame. For sufficiently small ε, the results of Section 4.8 show that periodic hyperbolic trajectories $\gamma_1(t)$ and $\gamma_2(t)$ exist within a distance of order ε from the original steady hyperbolic points. As in the preceding examples, the time-dependent nonlinear unstable and stable material manifolds $W^u[\gamma_1(t)]$ and $W^s[\gamma_2(t)]$ of $\gamma_1(t)$ and $\gamma_2(t)$ can be obtained using analytic solutions for their linear approximations $E^s[\gamma_1(t)]$ and $E^u[\gamma_2(t)]$, combined with numerical solutions using the methods outlined in Appendix C.

Consider now the time-dependent structure of the material manifolds $W^u[\gamma_1]$ and $W^s[\gamma_2]$. At a given time t, these manifolds consist of the sets of all points through which pass trajectories $\mathbf{x}(t)$ that are asymptotic to γ_1

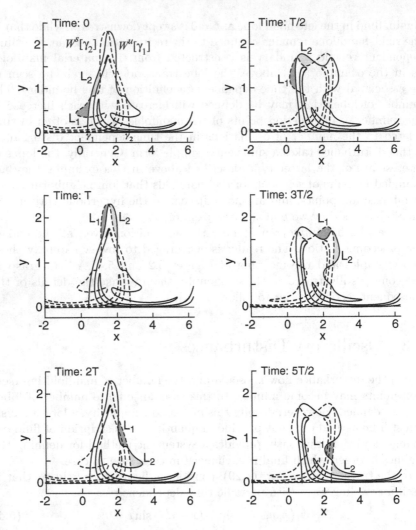

FIGURE 5.5. Material (unstable, solid line; stable, dashed) manifolds for the periodic disturbance $\psi_1 = \sin k_1(x - c_1' t) \sin l_1 y$ with $\varepsilon = 0.3$, $k_1 = 1, l_1 = 2$, and $\omega = k_1 c_1' = \pi$. The manifolds are shown at times that are multiples of $T/2$, where $T = 2$ is the period of the disturbance. The lobes L_1 (dark shading) and L_2 (light shading) are indicated. Only finite segments of the manifolds are shown, contiguous with the corresponding hyperbolic trajectories.

as $t \to -\infty$, or to γ_2 as $t \to \infty$, respectively. Although the velocity field is periodic in time, these individual asymptotic trajectories $\mathbf{x}(t)$ cannot be, since they tend toward specific limits rather than oscillating periodically. However, the periodicity of the velocity field does imply that each such trajectory $\mathbf{x}(t)$ will be followed by an exact duplicate $\mathbf{x}(t - T)$ that is delayed in time by one period T, and preceded by an exact duplicate $\mathbf{x}(t+T)$ that is advanced in time by one period T. Moreover, each such duplicate will in turn be followed and preceded by another such duplicate, and so on ad infinitum. Thus, the set of all points through which pass trajectories with a given asymptotic limit will, unlike the individual trajectories themselves, be periodic in T, since each such trajectory has an infinite set of identical duplicates that lag or lead by integer multiples of the period T. Consequently, for the time-periodic disturbance (5.4) with period T, the unstable and stable material manifolds $W^u[\gamma_1]$ and $W^s[\gamma_2]$ must be periodic functions of t, with period T (Figure 5.5). Note that these manifolds will not remain fixed during each period of oscillation, but will repeat precisely the same motion during the course of each period. The intermediate motion is not a standing oscillation, but a kind of progressive wave-like motion, in which the manifold structure distorts and rotates around the recirculation regime. This motion arranges itself, as it must, so that the result of the distortion and rotation is a final manifold structure that is congruent to the initial structure after each successive period T of the oscillation.

As in the preceding examples of transient disturbances, it is possible to construct lobe and regime boundaries from intersection points of material manifolds $W^u[\gamma_1]$ and $W^s[\gamma_2]$, and the segments of these manifolds that join the intersection points. A powerful and convenient simplification of this construction arises from the periodicity of the manifolds for the periodic disturbance (5.4). Namely, the regime boundaries may in this case be chosen to be fixed in space at times that are offset by integer multiples of the period T.

In the traveling wave example with the periodic disturbance (5.4), the manifold and lobe structure continuously oscillate between the two repeating forms reached at $t = 0$ and $t = T/2$ (for example, with $\varepsilon = 0.3, k_1 = 1, l_1 = 2$, and $\omega = \pi$, as in Figure 5.5). Fluid initially in adjacent lobes on the upstream edge of the recirculation regime moves around the recirculation and toward the second hyperbolic trajectory as time evolves. Ultimately, the adjacent lobes separate just as in the conceptual example (Figure 5.1), with fluid in one lobe recirculating and that in the other moving downstream. Note that, as anticipated from the arguments above, the manifold and lobe structure repeats after each complete period T. Since this structure is fixed, when regarded at a sequence of times offset by integer multiples of the period T, the exchange of fluid between the recirculation and the jet regimes may be completely analyzed in terms of the fixed structure of the lobes, regarded at this same sequence of times. This perspective is developed in quantitative detail in the next chapter.

When the oscillatory velocity is quasiperiodic, that is, composed of two or more incommensurate frequencies, the manifold structure is no longer exactly periodic. It is still possible, however, to find hyperbolic trajectories and their unstable and stable manifolds, compute these and their intersections, define lobes, and analyze the fluid exchange process a similar manner.

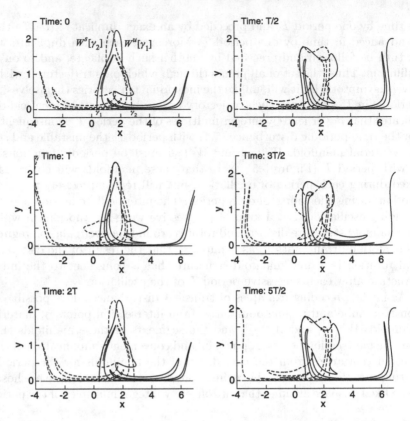

FIGURE 5.6. Material (unstable, solid line; stable, dashed) manifolds for the quasiperiodic disturbance (5.5) for $\varepsilon = 0.3$, $g_1 = 0.99$, $g_2 = 0.141$, $\omega = [\pi, \pi/2]$, and $k_i = 1$, $l_i = 2$, $i = 1, 2$. The manifolds are shown at multiples of $T/2 = 1$, where $T = 2$. Only finite segments of the manifolds are shown, contiguous with the corresponding hyperbolic trajectories.

A general quasiperiodic disturbance velocity for the traveling wave example can be expressed in the form

$$\psi_1(x, y, t) = \sum_{i=1}^{n} g_i \sin(k_i x - \omega_i t) \sin l_i y \qquad (5.5)$$

where $\omega_i = k_i c_i' = k_i(c_0 - c_i)$ are the disturbance frequencies in the co-moving frame, (k_i, l_i) are the disturbance wavenumbers, and g_i are the respective disturbance amplitudes, for $i = 1, \ldots, n$. For a representative case of two-frequency $(n = 2)$ quasiperiodic time-dependence (for example, $\varepsilon = 0.3, g_1 = 0.99, g_2 = 0.141, \omega_1 = \pi, \omega_2 = \pi/2, k_1 = k_2 = 1, l_1 = l_2 = 1$, as in Figure 5.6), qualitative similarities to the periodic case (5.4) are evident, such as approximate symmetries, general appearance of the lobe structure, characteristic movement patterns of the lobes, and other features, but notable differences arise also. For a given amplitude of the disturbance, the geometry of the lobes is more complex for the quasiperiodic case than for the periodic case. Several new lobes of smaller area appear in the quasiperiodic case that are not found in the periodic case. The number and the areas of such lobes are highly dependent on individual frequency values and the relative amplitudes of the various components of the quasiperiodic disturbance. In this quasiperiodic example, the lobe geometry approximately repeats itself at the largest period $(2T)$ of the disturbance velocity field (for example, at $t = 0$ and $t = 2T$, or at $t = T/2$ and $5T/2$ in Figure 5.6). As in the periodic case, but now only approximately, the lobe geometry also oscillates on a shorter period between two extreme forms (for example, those at $t = 0$ and $t = T/2$).

5.4 Summary

When the traveling wave becomes time-dependent in the translating frame, exchange of fluid between flow regimes is generally induced. The associated Lagrangian motion can be characterized by identifying and analyzing a special set of material curves, the stable and unstable material manifolds of hyperbolic trajectories. It is this set of material curves that were sketched in the illustrations of the breakup of regime boundaries in Chapter 1 (Figures 1.2 and 1.5). Intersections of the manifolds are points through which pass special, doubly asymptotic trajectories. These intersections and the segments of material manifolds that join them define the boundaries of *lobes*, patches of fluid that pass from one fluid regime to another in the time-dependent flow. Analysis of the evolution of the material manifolds and lobes provides a complete, quantitative picture of the exchange of fluid between flow regimes. In this picture, exchange of fluid between flow regimes is always associated with changes in the definition of the regime boundaries.

For the traveling wave, and for other similar flows, there are important, characteristic differences between the cases of transient, periodic, and quasiperiodic time-dependence in the disturbance velocity field. For transient disturbances, material manifolds may intersect at only a finite number of points, and form only a finite number of lobes. For periodic or quasiperiodic disturbances, material manifolds may intersect at an infinite number of points. For periodic disturbances, the material manifold structure is also exactly periodic, and this periodicity may be exploited to simplify the corresponding analysis. For quasiperiodic disturbances, approximate periodicity allows a related construction under certain circumstances. For transient disturbances, such a construction is not possible. The geometry of the material manifolds and the corresponding lobe structure and fluid exchange processes depend strongly on the details of the velocity field, whatever the underlying time-dependence.

This analysis has focused on the fluid that is transported between regimes, which is contained in lobes defined by the intersections of material manifolds. Fluid that is not contained in these lobes will remain within its original regime. In two-dimensional velocity fields with periodic or quasiperiodic time-dependence, fluid motions in these other regions are often themselves periodic or quasiperiodic, and their trajectories may be identified as, or enclosed by, KAM tori (Chapter 3). For the transient disturbances (5.1) and (5.3), for example, there are regions of fluid in the recirculating regime that are not contained within any lobes, and around which the manifolds wrap as they move with the flow (Figures 5.2, 5.4). The boundaries of these areas exterior to the lobes are segments of the same material manifolds that form the boundaries of the lobes. Like KAM tori in oscillatory flows, these regions can act as barriers to fluid transport in transient flows.

5.5 Notes

The transport by a steadily translating nonlinear wave of fluid enclosed by a closed streamline in a reference frame moving with the wave is a familiar concept in fluid mechanics. The generalization of this concept to time-dependent perturbations of the traveling wave leads naturally to the consideration of the stable and unstable material manifolds described in this chapter. For reasons discussed by Beigie et al. (1994), stable and unstable manifolds of hyperbolic trajectories form coherent structures in a wide variety of flows, in an engineering fluid mechanics context such as studied by Shariff et al. (1992) and Duan & Wiggins (1997) as well as the geophysical fluid flows of primary interest here. The role of analogous structures, in phase space rather than physical space, for the nonlinear stability of dynamical systems was first recognized by Poincaré (1892) and has since received much attention in work such as that of Melnikov (1963); see, for example, the dynamical systems texts by Guckenheimer & Holmes (1983), Wiggins (1988), and Wiggins (2003). Numerical methods for computing hyperbolic trajectories and their stable and unstable manifolds are discussed below in Appendix C.

6

Lobe Transport and Flux

6.1   Regime and Lobe Boundaries

In Chapter 5, a lobe-transport analysis of fluid exchange in the traveling-wave flow was conducted for a simple transient pulse disturbance, and a similar approach was outlined for disturbances with general, oscillatory time-dependence. In this chapter, the quantitative analysis is carried out for these more complex disturbances, and a method is developed that extends and generalizes the approach.

A crucial element of these constructions is the identification of the time-dependent Lagrangian regime boundaries, across which the lobe transport occurs. This identification is perhaps most clearly motivated in the case of time-periodic disturbances, for which the relevant regime boundaries are, from a specific point of view and for reasons outlined in the previous chapter, fixed in space. This case is therefore discussed first below, followed by the more general development for disturbances with arbitrary time-dependence.

The identification of these regime boundaries is a necessary step in the lobe-transport analysis. It also represents a critical link with the physical problem, on which the practical significance and utility of the subsequent analysis hinge. In a loose sense, one can anticipate that a useful identification will be possible in strongly inhomogeneous flows that are dominated by coherent features, such as the time-dependent traveling waves and meandering jets that serve as the primary example here. For such flows, the very existence of the coherent features suggests that regime boundaries may be constructed that will serve as quantitative characterizations of these same features. Whether this general approach will prove appropriate for flows that are statistically homogeneous, or approximately so, and in which spatially coherent features are either strongly transient or difficult to identify at all, remains an open question.

Regime boundaries and lobes for the traveling-wave flow with transient pulse disturbance (5.1) were identified in Chapter 5 (Section 5.2). The next section extends this analysis to the case of the time-periodic disturbance (5.4). In preparation for this extension, which includes the definition and interpretation of several deceptively simple geometric objects and illustrations, an admonition given twice in that chapter is repeated once more here: a solid grasp of the meaning of the illustrations and constructions in that previous example should make the material in this chapter easier to follow.

6.2 The Traveling Wave with Time-Periodic Disturbance

For the traveling wave with time-periodic wave disturbance (5.4), the hyperbolic trajectories $\gamma_1(t)$ and $\gamma_2(t)$ and their unstable and stable material manifolds $W^u[\gamma_1(t)]$ and $W^s[\gamma_2(t)]$ were computed in Section 5.3. The trajectories $\gamma_1(t)$ and $\gamma_2(t)$ are periodic, and it was shown in Chapter 5 that, for the case of a time-periodic disturbance, the material manifolds $W^u[\gamma_1(t)]$ and $W^s[\gamma_2(t)]$ are also periodic. This allows the lobe-transport analysis to be conducted in terms of a regime boundary that is fixed in space.

The periodicity of the material manifolds was illustrated in Figure 5.5, which shows that the structure of these manifolds repeats exactly after each period $T = 2\pi/(k_1 c_1')$. The fixed regime boundaries are constructed by restricting attention to one of these repeating patterns, at the sequence of times $\mathcal{S} = \{t + jT, \ j = 0, \pm1, \pm2, \pm3, \dots\}$. A convenient choice for this sequence for the traveling wave with disturbance (5.4) is $t = 0$, so that $\mathcal{S} = \mathcal{S}_1 = \{jT, \ j = 0, \pm1, \pm2, \pm3, \dots\}$, at which times the manifold structure is symmetric in x about the midpoint of the recirculation regime (Figure 5.5). The regime boundary $\mathcal{B}$ is constructed from the segments of $W^u[\gamma_1]$ and $W^s[\gamma_2]$ that join the hyperbolic trajectory points $\gamma_1(t = 0)$ and $\gamma_2(t = 0)$, respectively, with the manifold intersection point $\mathbf{q}$, where the x-coordinate of $\mathbf{q}$ is the midpoint in x between γ_1 and γ_2 (Figure 6.1). At each time in the sequence $\mathcal{S}_1$, this boundary $\mathcal{B}$ will have exactly the same spatial structure, because of the periodicity of the material manifolds. Thus, it may be regarded as fixed in space, provided that attention is restricted to the sequence of times $\mathcal{S}_1$. There is a technical restriction on the manifold intersection point $\mathbf{q}$, which this choice satisfies. Namely, $\mathbf{q}$ must be a *primary intersection point*: the two segments of $W^u[\gamma_1]$ and $W^s[\gamma_2]$ that meet at $\mathbf{q}$ and together constitute $\mathcal{B}$ must not intersect at any other point. An additional technical assumption made in this analysis, which is always satisfied for two-dimensional fluid flow, is that the flow is *orientation-preserving*: a closed material curve is deformed by the flow in such a manner that the fluid elements along the curve are always met in the same order if the curve is traversed in a fixed (say, counterclockwise) direction.

It is important to recognize that, while the regime boundary $\mathcal{B}$ has been constructed so that it is fixed in space and identical at each time in the sequence $\mathcal{S}$, the trajectories that pass through the points of $\mathcal{B}$ represent moving fluid elements, and are not fixed in space. In particular,

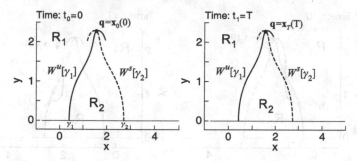

FIGURE 6.1. The boundary $\mathcal{B}$ between the jet (R_1) and recirculation (R_2) flow regimes for the traveling wave with time-periodic wave disturbance is constructed from segments of the unstable (solid line) and stable (dashed line) material manifolds of the hyperbolic trajectories γ_1 and γ_2. The boundary $\mathcal{B}$ is identical at $t = t_0 = 0$ and $t = t_1 = T$ (and at all $t_j = jT$, $j = 0, \pm 1, \pm 2, \pm 3, \ldots$), but two different trajectories, $\mathbf{x}_0(t)$ and $\mathbf{x}_T(t)$, pass through the manifold intersection point $\mathbf{q}$ at $t = t_0 = 0$ and $t = t_1 = T$, respectively.

the trajectory $\mathbf{x}_T(t)$ that passes through the point $\mathbf{q}$ at $t = T$ is not the same trajectory $\mathbf{x}_0(t)$ that passed through $\mathbf{q}$ at $t = 0$. Instead, the former is an exact duplicate of the latter, lagged in time by the period T: $\mathbf{x}_T(t) = \mathbf{x}_0(t - T)$. Thus, a different trajectory passes through the point $\mathbf{q}$ at each time in the sequence $\mathcal{S}_1$. The same is true of every point of the boundary $\mathcal{B}$. Note that $\mathcal{S}_1$ is an infinite sequence. Thus, at each time t, there are an infinite number of manifold intersection points, located at the points $\mathbf{x} = \mathbf{x}_j(t)$, $j = 0, \pm 1, \pm 2, \pm 3, \ldots$. At any time t, most of these points are confined to small regions around either $\gamma_1(t)$ or $\gamma_2(t)$, to which the solutions $\mathbf{x}_j(t)$ are all asymptotic as $t \to -\infty$ and $t \to +\infty$, respectively.

Consider now the two trajectories $\mathbf{x}_0(t)$ and $\mathbf{x}_T(t)$, which pass through $\mathbf{q}$ at $t = 0$ and $t = T$, respectively, along with the boundary $\mathcal{B}$ and the segment of $W^s[\gamma_2]$ between $\mathbf{x}_0(t)$ and $\mathbf{x}_T(t)$ (Figure 6.2). At $t = 0$, the latter segment is not part of $\mathcal{B}$, which includes only the portion of $W^s[\gamma_2]$ between $\mathbf{q}$ and γ_2. Instead, it and the segment of $\mathcal{B}$ between $\mathbf{x}_0(0)$ and $\mathbf{x}_T(0)$ form the boundaries of two enclosed regions. These two regions are the lobes that will be used to quantify transport into and out of the flow regime bounded by $\mathcal{B}$ (Figure 6.2). The boundaries of the two lobes meet at a single point, the manifold intersection point $\mathbf{p}$ that, at any given time t, lies between the intersection points $\mathbf{x}_0(t)$ and $\mathbf{x}_T(t)$.

As the flow evolves from $t = 0$, when $\mathbf{x}_0(t = 0) = \mathbf{q}$, through one oscillation period to $t = T$, when $\mathbf{x}_T(t = T) = \mathbf{q}$, these two lobes move from the upstream side of $\mathbf{q}$ to the downstream side of $\mathbf{q}$ (Figure 6.2). In so doing, they also cross the regime boundary $\mathcal{B}$. At $t = 0$, the lobe adjacent to $\mathbf{x}_T$ is in the interior of the recirculation region enclosed by $\mathcal{B}$, denoted by R_2, and the lobe adjacent to $\mathbf{x}_0$ is in the exterior jet region, denoted by R_1. At $t = T$, the lobe adjacent to $\mathbf{x}_T$ has crossed $\mathcal{B}$ to the exterior jet region

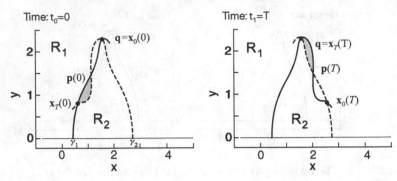

FIGURE 6.2. Turnstile lobes governing transport between flow regimes R_1 and R_2 at the initial time $t_0 = 0$ and the final time $t_1 = T$, for the traveling wave with time-periodic disturbance. The boundaries of the flow regimes are the segments of material manifolds that intersect at $\mathbf{q}$, as indicated in Figure 6.1. The trajectories $\mathbf{x}_0(t)$ and $\mathbf{x}_T(t)$ pass through the manifold intersection point $\mathbf{q}$ at $t = t_0$ and $t = t_1$, respectively. The locations of the points $\mathbf{x}_0(t_1)$ and $\mathbf{x}_T(t_0)$, on the opposite boundary of the lobes from $\mathbf{q}$, are also indicated, as are the intermediate manifold-intersection points $\mathbf{p}(t_0)$, $\mathbf{p}(t_1)$ on the trajectory $\mathbf{p}(t)$. The shaded turnstile lobe is in the interior recirculation regime R_2 at t_0, and in the exterior jet regime R_1 at t_1, while the unshaded turnstile lobe is in the exterior jet regime R_1 at t_0, and in the interior recirculation regime R_2 at t_1..

(R_1), and the lobe adjacent to $\mathbf{x}_0$ has crossed $\mathcal{B}$ to the interior recirculation region (R_2). The lobes may be identified by these regime transitions: $L_{1,2}$ denotes the lobe that moves from region R_1 to region R_2, while $L_{2,1}$ denotes the lobe that moves from region R_2 to region R_1.

Note that, as in Chapter 5, fluid exchange of this type can occur only when the regime boundaries are redefined. Although the boundary $\mathcal{B}$ is fixed in space, it is composed of different material elements at different times. At $t = 0$, $\mathcal{B}$ can be decomposed into two segments of $W^u[\gamma_1]$, divided by the point $\mathbf{x}_T(0)$, and the single segment of $W^s[\gamma_2]$ described above. At $t = T$, the segment of $W^u[\gamma_1]$ between γ_1 and $\mathbf{x}_T$ has been stretched out by the flow, and now forms the entire upstream half of $\mathcal{B}$, from γ_1 to $\mathbf{q}$. Now, at $t = T$, it is the downstream half of $\mathcal{B}$ that can be decomposed, into two segments of $W^s[\gamma_2]$ divided by the point $\mathbf{x}_0(T)$. Thus, in this decomposition the two segments of $\mathcal{B}$ that are adjacent to the hyperbolic trajectories γ_1 and γ_2 are distorted by the flow during the interval from $t = 0$ to $t = T$, but remain contiguous. The portion of $\mathcal{B}$ between $\mathbf{x}_0$ and $\mathbf{x}_T$, however, is redefined during this interval: at $t = 0$, it is a segment of $W^u[\gamma_1]$, while at $t = T$, it is a segment of $W^s[\gamma_2]$. It is this material redefinition of the regime boundary $\mathcal{B}$ that allows the fluid exchange to occur. No exchange can occur along the segments of $\mathcal{B}$ adjacent to γ_1 and γ_2, since these remain contiguous, and consequently all of the fluid exchange is mediated by the two lobes. In this sense, the two lobes might be said to control or determine the rate of fluid exchange, in a manner that is reminiscent of turnstiles, and consequently they are often referred to as *turnstile lobes*.

The area of each turnstile lobe, which must remain constant in the incompressible flow, determines the volume of fluid transported into or out of the recirculation regime defined by $\mathcal{B}$ during the time interval from $t = 0$ to $t = T$. Since an exactly equivalent movement of lobes into and out of $\mathcal{B}$ occurs as the periodic flow evolves from each point in the sequence of times $\mathcal{S}$ to each successive point, T time units later, the movement of the turnstile lobes provides a complete, quantitative description of the exchange of fluid between the recirculation and jet regimes, as defined by the boundary $\mathcal{B}$. The average rate of fluid exchange can be computed directly from the areas of these two lobes and the period T.

6.3   General Oscillatory Disturbances

Suppose now that the traveling wave disturbance is oscillatory, but not periodic with fixed period T. For example, it might be quasiperiodic, or even aperiodic or chaotic, with essentially random oscillatory behavior. In this more general case, it is still possible to carry out a lobe-transport analysis similar to that in the previous section, provided that the disturbances are not so large or extreme that recirculation and jet regimes cannot be identified in a similar way.

The first step in this generalized analysis is to identify the relevant hyperbolic trajectories $\gamma_1(t)$ and $\gamma_2(t)$, which take the place of the hyperbolic periodic trajectories for the periodic disturbance. Since the disturbances are not periodic, these hyperbolic trajectories will not be periodic, as they were in the preceding section. However, if the disturbance is not too large, identification of hyperbolic trajectories of the form (4.55) near the steady-flow stagnation points will still be possible. The second step is then to construct the corresponding stable and unstable material manifolds of $\gamma_1(t)$ and $\gamma_2(t)$, as outlined in Chapter 4.

For the traveling wave with quasiperiodic disturbance, these two steps were carried out for a particular set of parameters in Chapter 5. The structure of the resulting manifolds for the quasiperiodic disturbance (Figure 5.6) is qualitatively similar to the manifold structure for the periodic disturbance (Figure 5.5). In both cases, the contorted, complex, looping curves of the manifolds appear confined to the transition region between the recirculation and jet regimes. This confinement suggests that they might be used as quantitative representations of the regime boundary. In the preceding section, this was done for the periodic case, and a similar construction is possible for the quasiperiodic case. The main complication is that the regime boundary $\mathcal{B}$ is no longer fixed in space as time evolves, because the manifold structure no longer repeats exactly.

The third step, then, is to identify a sequence $\mathcal{S} = \{\ldots, t_{-1}, t_0, t_1, \ldots\}$

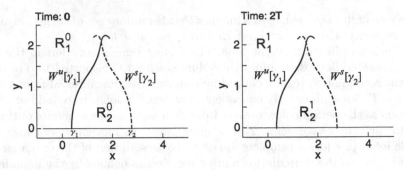

FIGURE 6.3. Boundaries $\mathcal{B}(t_0)$ and $\mathcal{B}(t_1)$ between the jet and recirculation flow regimes for the traveling wave with quasiperiodic wave disturbance, constructed from segments of the unstable (solid line) and stable
(dashed line) material manifolds of the hyperbolic trajectories γ_1 and γ_2. Here
$\varepsilon = 0.3$, $g_1 = 0.99$, $g_2 = 0.141, \omega = [\pi, \pi/2]$, and $k_i = 1$, $l_i = 2$, $i = 1, 2$. The
manifolds are shown at the two times $t_j = jT/2$, $j = \{0, 1\}$, where $T = 2$, for
which the manifold structure is nearly, but not exactly, periodic. The exterior jet
regimes are R_1^0 at t_0 and R_1^1 at t_1, and the interior recirculation regimes are R_2^0
at t_0 and R_2^1 at t_1.

of times at which the time-dependent regime boundary $\mathcal{B}(t_j)$ is to be constructed from the material manifolds $W^u[\gamma_1]$ and $W^s[\gamma_2]$, and to construct
$\mathcal{B}(t_j)$ at these times $\mathcal{S}$. This generalizes the regular sequence $\mathcal{S}_1$ that was
used in the periodic case, since the times in the sequence $\mathcal{S}$ need not be
separated by a fixed interval T, as were the times in the sequence $\mathcal{S}_1$. For
quasiperiodic disturbances, this sequence will again always be infinite, as it
was in the periodic case. For the specific quasiperiodic example of Chapter
5 (Figure 5.6), the manifold structure does approximately repeat after each
fixed interval $2T$, so in this case the choice $\mathcal{S} = \mathcal{S}_2 = \{\ldots, -2T, 0, 2T, \ldots\}$
is appropriate (Figure 6.3). A boundary $\mathcal{B}(t_j)$ can be constructed that is
approximately symmetric again at each time t_j (Figure 6.3), using segments of the material manifolds that intersect at a single point $\mathbf{q}_j$. The
intersection points $\mathbf{q}_j$ and the boundary $\mathcal{B}(t_j)$ are now time-dependent,
rather than fixed in space as in the periodic case. The regions interior and
exterior to the time-dependent boundary $\mathcal{B}(t_j)$ are the time-dependent jet
and recirculation regimes $R_1^j = R_1(t_j)$ and $R_2^j = R_2(t_j)$, respectively.

The fourth and final step is to identify the turnstile lobes associated
with the time-dependent boundary $\mathcal{B}(t_j)$, in the same manner as for the
periodic case. Since the boundary is now time-dependent, this must now
be done separately for each time t_j in $\mathcal{S}$. For example, the trajectory $\mathbf{x}_0(t)$
that passes through the intersection point $\mathbf{q}_0$ (so that $\mathbf{x}_0(t_0) = \mathbf{q}_0$) can
be traced forward to $t = t_1$, and the trajectory $\mathbf{x}_1(t)$ that passes through
the intersection point $\mathbf{q}_1$ (so that $\mathbf{x}_1(t_1) = \mathbf{q}_1$) can be traced backward to
$t = t_0$. For this step, a necessary technical condition is that $\mathbf{x}_1(t_0)$ be closer

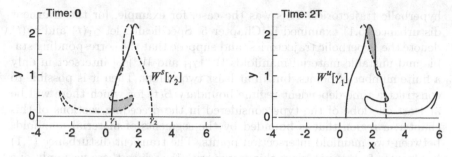

FIGURE 6.4. Turnstile lobes governing transport between the jet and recirculation flow regimes between times $t_0 = 0$ and $t_1 = 2T$, for the traveling wave with
quasiperiodic disturbance. Here $\varepsilon = 0.3$, $g_1 = 0.99$, $g_2 = 0.141$, $\omega = [\pi, \pi/2]$,
and $k_i = 1$, $l_i = 2$, $i = 1, 2$. The lobes and regime boundaries are shown at the
two times $t_j = j2T$, $j = \{0, 1\}$, where $T = 2$, for which the manifold structure
is nearly, but not exactly, periodic. The shaded turnstile lobe areas are in the
interior recirculation regime $R_2^0 = R_2(t_0)$ at t_0, and in the exterior jet regime
$R_1^1 = R_1(t_1)$ at t_1, while the unshaded turnstile lobe areas are in the exterior jet
regime $R_1^0 = R_1(t_0)$ at t_0, and in the interior recirculation regime $R_2^1 = R_2(t_1)$ at
t_1. Note that the shaded and unshaded turnstile lobes are each made up of two
separate, unconnected lobe areas, one larger and one smaller.

to the hyperbolic trajectory γ_1, in the sense of arc length measured along
the unstable manifold $W^u[\gamma_1]$, than $\mathbf{x}_0(t_0)$. Denote by $L_{1,2}^j$ and $L_{2,1}^j$ the
turnstile lobes that describe the exchange between the time-dependent jet
and recirculation regimes $R_1(t_j)$ and $R_2(t_{j+1})$, and $R_2(t_j)$ and $R_1(t_{j+1})$,
respectively, during the interval $t = t_j$ to $t = t_{j+1}$. These lobes are the
regions enclosed by the segments of the material manifolds between $\mathbf{x}_j(t)$
and $\mathbf{x}_{j+1}(t)$. For example, $L_{1,2}^0$ and $L_{2,1}^0$ are the regions enclosed by the
segments of the material manifolds between $\mathbf{x}_0(t)$ and $\mathbf{x}_1(t)$ (Figure 6.4).
Note that, for the quasiperiodic flow of this example, the turnstile lobes
$L_{1,2}^0$ and $L_{2,1}^0$ are each made up of two separate, unconnected lobe areas,
one larger and one smaller. When multiple, unconnected lobe areas cross
the regime boundary during a single step in the time sequence $\mathcal{S}$, as they
do in this case, it is convenient still to refer in this way to them jointly as
a single *turnstile lobe*. Recall also that, since the flow is incompressible, the
area of a given lobe region cannot change with time, despite the changes
in the shapes of the lobes.

6.4 Transient Disturbances

Suppose that the velocity field consists of the steady part plus a transient
disturbance, and that there are consequently only a finite number of intersection points of the unstable and stable material manifolds of the relevant

hyperbolic trajectories. This was the case, for example, for the transient disturbance (5.1) examined in Chapter 5. Specifically, let $\gamma_1(t)$ and $\gamma_2(t)$ denote the hyperbolic trajectories, and suppose that the corresponding stable and unstable material manifolds $W^u[\gamma_1]$ and $W^s[\gamma_2]$ intersect in only a finite number of points, but in at least two points. Then it is possible to construct a time-dependent regime boundary $\mathcal{B}(t)$ for which there will be at least one lobe of the type considered in the preceding sections of this chapter: a region that is bounded by the segments of material manifolds between two manifold intersection points. The transient disturbance (5.1) in Chapter 5 does not meet this latter criterion, since there was only one such manifold intersection point, and the lobe boundaries included the hyperbolic trajectories γ_1 and γ_2 themselves. That special case can generally be handled by following the discussion of the example (5.1) in Chapter 5.

The lobe analysis in this case proceeds essentially as in the case of oscillatory disturbances with an infinite number of manifold intersection points, except that the sequence $\mathcal{S}$ will be finite. The first and second steps of the analysis are to identify the relevant hyperbolic trajectories $\gamma_1(t)$ and $\gamma_2(t)$, and to construct the corresponding stable and unstable material manifolds of $\gamma_1(t)$ and $\gamma_2(t)$, as before. The third step, then, is to identify a finite sequence $\mathcal{S} = \{t_{-J}, t_{-J+1}, \ldots, t_{-1}, t_0, t_1, \ldots, t_{K-1}, t_K\}$ of $J + K + 1$ times at which the time-dependent regime boundary $\mathcal{B}(t_j)$ is to be constructed from the material manifolds $W^u[\gamma_1]$ and $W^s[\gamma_2]$, and to construct $\mathcal{B}(t_j)$ at these times $\mathcal{S}$. This construction proceeds exactly as before, except that $\mathcal{S}$ is now finite. The fourth and final step is to identify the turnstile lobes associated with the time-dependent boundary $\mathcal{B}(t_j)$. Since the boundary is again time-dependent, this must again be done separately for each time t_j in $\mathcal{S}$. An important distinction of the case of transient disturbances is that for $t > t_K$, where t_K is the last time in the finite sequence $\mathcal{S}$, the exchange process is completed, and terminates. No additional transport can occur, since all of the turnstile lobes have by then passed the last manifold intersection point $\mathbf{q}_K$ (Figure 6.5). For the traveling wave, disturbances of this type are discussed in Section 5.2 (Figures 5.2, 5.4).

6.5   Lobe Area and Flux Formulas

For two-dimensional, incompressible, time-dependent flows, there is an integral formula for the area of a lobe that involves, remarkably, only the fluid trajectories passing through the two intersection points of the material manifolds that enclose the lobe. Denote these two fluid trajectories by $\mathbf{x}_p(t)$ and $\mathbf{x}_q(t)$. Then the area $A(L^j)$ of the lobe L^j (where, for example, L^j might represent either $L^j_{1,2}$ or $L^j_{2,1}$) is given by

$$A(L^j) = \int_{\mathbf{x}_p} (y\, dx + \psi\, dt) - \int_{\mathbf{x}_q} (y\, dx + \psi\, dt), \tag{6.1}$$

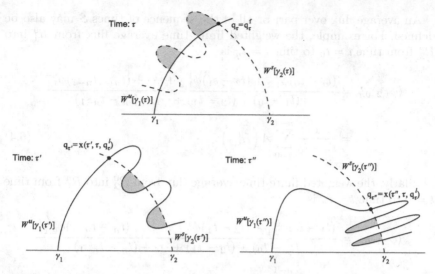

FIGURE 6.5. Transport in the case of a finite number of lobes (but more than one). The hyperbolic trajectories and their unstable (solid line) and stable (dashed line) manifolds are shown for three times: $t = \tau < \tau' < \tau''$.

where $\psi(x, y, t)$ is the streamfunction for the velocity field, and the integrals are line integrals along the trajectories $\mathbf{x}_p(t)$ and $\mathbf{x}_q(t)$ over the infinite time interval $-\infty < t < \infty$ (Kaper & Wiggins (1991), Kaper & Wiggins (1992)). For example, for the turnstile lobe $L_{1,2}^j$, $\mathbf{x}_p(t) = \mathbf{x}_{j+1}(t)$, and $\mathbf{x}_q(t)$ is the trajectory through the other manifold intersection point on the boundary of $L_{1,2}^j$, which is located between $\mathbf{x}_{j+1}$ and $\mathbf{x}_j$ along $W^u[\gamma_1]$ and $W^s[\gamma_2]$. Similarly, for $L_{2,1}^j$, the two trajectories are $\mathbf{x}_p(t) = \mathbf{x}_j(t)$ and the same $\mathbf{x}_q(t)$ as for $L_{1,2}^j$. The absolute value $|A(L^j)|$ can be taken to guarantee positivity of the area. The formula (6.1) is exact, and requires no additional restrictions on the trajectories $\mathbf{x}_p(t)$ and $\mathbf{x}_q(t)$. However, its use in general situations will often require numerical calculations.

Since the fluid moving from region to region is completely contained in the turnstile lobes, the corresponding fluxes can be computed directly from the lobe areas. Then the average flux across $\mathcal{B}(t_j)$ from R_1^j into R_2^{j+1} during the interval $t_j < t < t_{j+1}$ may be defined as the rate

$$\phi_{1,2}^j = \frac{1}{t_{j+1} - t_j} A\left(L_{1,2}^j\right). \qquad (6.2)$$

Similarly, the average flux across $\mathcal{B}(t_j)$ from R_2^j into R_1^{j+1} during the interval $t_j < t < t_{j+1}$ may be defined as the rate

$$\phi_{2,1}^n = \frac{1}{t_{j+1} - t_j} A\left(L_{2,1}^j\right). \qquad (6.3)$$

An average flux over part or all of the sequence of times S may also be defined. For example, the weighted finite-time average flux from R_1^0 into R_2^n from time $t = t_0$ to time $t = t_n$ is

$$\langle \phi_{1,2} \rangle_{0,n} = \frac{(t_1 - t_0)\phi_{1,2}^0 + (t_2 - t_1)\phi_{1,2}^1 + \cdots + (t_n - t_{n-1})\phi_{1,2}^{n-1}}{(t_1 - t_0) + (t_2 - t_1) + \cdots + (t_n - t_{n-1})}$$

$$= \frac{1}{t_n - t_0} \sum_{j=0}^{n-1} A\left(L_{1,2}^j\right). \tag{6.4}$$

Similarly, the weighted finite-time average flux from R_2^0 into R_1^n from time $t = 0$ to time $t = t_n$ is

$$\langle \phi_{2,1} \rangle_{0,n} = \frac{(t_1 - t_0)\phi_{2,1}^0 + (t_2 - t_1)\phi_{2,1}^1 + \cdots + (t_n - t_{n-1})\phi_{2,1}^{n-1}}{(t_1 - t_0) + (t_2 - t_1) + \cdots + (t_n - t_{n-1})}$$

$$= \frac{1}{t_n - t_0} \sum_{j=0}^{n-1} A\left(L_{2,1}^j\right). \tag{6.5}$$

These fluxes, computed from the lobe areas for the full time sequence S, provide a complete quantitative description of the exchange of fluid between the regions R_1^j and R_2^j.

6.6 Notes

In the context of area-preserving maps, the turnstile mechanism of transport was first noted by Bartlett (1982) and Channon & Lebowitz (1980). The turnstile mechanism, along with an assumption on chaotic trajectory behavior, was used to develop a Markov model for transport by MacKay et al. (1984). Lobe dynamics for two-dimensional maps was developed in Rom-Kedar et al. (1990) and Rom-Kedar & Wiggins (1990). This was later extended to two-dimensional quasiperiodically time-dependent velocity fields in Beigie et al. (1991) and to two-dimensional velocity fields with arbitrary time-dependence in Malhotra & Wiggins (1998). Formulas for the area of lobes are given in Kaper & Wiggins (1991) and Kovacic (1991). A numerical approach to computing lobe areas is described in Coulliette & Wiggins (2001). Lobe dynamics is applied to transport in a wind-driven double-gyre in Coulliette & Wiggins (2001), a barotropic meandering jet in Miller et al. (1997) and Rogerson et al. (1999), and to transport in a cellular flow in Camassa & Wiggins (1991). One possible way of extending lobe dynamics to three-dimensional velocity fields is described in Mezic & Wiggins (1994).

7

Transport and Dynamics

7.1 Dynamics of Geophysical Flows

The preceding chapters have focused on a kinematic description of Lagrangian motion in simple models of traveling waves and meandering jets. This focus has allowed the most direct and straightforward development of the basic aspects of the geometric mathematical approach to the analysis of Lagrangian motion. Explicit consideration of the dynamical processes and balances that generate and maintain such flows and features in geophysical fluids has thus far been avoided.

This chapter begins to connect this new Lagrangian approach with fundamental dynamical principles for geophysical fluids. The dynamical and physical interpretations of the results from the new Lagrangian perspective are areas of active current research, and consequently this chapter must serve only as a starting point, which the interested reader may use to begin a more systematic study of these topics. The purpose here is to give an accessible overview of some of the new insights into geophysical fluid flows that have recently been obtained using these methods.

The dynamical models of jets and waves considered here are all based on the *quasigeostrophic* approximation, which is appropriate for many large-scale, low-frequency flows in the ocean and atmosphere (e.g., Holton (1992), Pedlosky (1987), Salmon (1998)). These models have an evolution equation of the form

$$\frac{\partial q}{\partial t} + \frac{\partial \psi}{\partial x}\frac{\partial q}{\partial y} - \frac{\partial \psi}{\partial y}\frac{\partial q}{\partial x} = 0, \qquad (7.1)$$

where ψ is the geostrophic streamfunction for the horizontal motion field and q is the quasigeostrophic *potential vorticity*. In the simplest case of barotropic (i.e., constant density) flow on scales small compared to the Earth's radius (see, e.g., Pedlosky, 1987), q and ψ are related by the same Poisson equation that connects the vorticity and streamfunction of a two-dimensional fluid in a nonrotating frame:

$$\nabla^2 \psi(x, y) = q(x, y). \qquad (7.2)$$

For *baroclinic* (i.e., stratified) flow, the functions q, ψ, and the elliptic operator depend also on the vertical coordinate z, while the streamfunction ψ continues to define the geostrophic horizontal velocity at each fixed z.

Baroclinic flows may also be represented by a vertical stack of discrete horizontal layers, each of which is governed by an equation like (7.1); in that case, the layers are coupled through the corresponding elliptic partial differential equations that relate the layer potential vorticities to the streamfunctions.

Many of the first studies of fluid exchange and transport in meandering jets and traveling waves, including some of those discussed in the preceding chapters, were based on kinematic models that were motivated either by observations and heuristic arguments, or by an appeal to a linearized version of the dynamical equations. As shown above in Chapters 2 and 4, the linearized motion field often provides a qualitatively correct description of the local Lagrangian motion near a point or trajectory, but in general does not represent the nonlinear partition of the domain into Lagrangian flow regimes, or the nonlinear trajectories associated with exchange between regimes. Since the nonlinearity in (7.1) is associated precisely with the Lagrangian motion, through the advection of the potential vorticity, linearization also necessarily removes essential elements of the dynamical balances. Consequently, none of these kinematic models conserved potential vorticity, or any related dynamical quantities, along fluid trajectories. Since such conservation properties are fundamental to the dynamics of many oceanic and atmospheric flows, including those that were the intended subject of the kinematic models, the question naturally arises whether the characteristic structures and mechanisms of fluid exchange exhibited in the kinematic models, including hyperbolic trajectories, material manifolds, and lobe transport, will indeed be found in dynamical models.

A related question concerns the integrability of two-dimensional flow (7.1) that conserves a vorticity or potential vorticity q. This is a special case of the situation discussed in Section 3.7, since (7.1) can be considered a form of the scalar advection equation. The scalar integral q is in this case functionally related to ψ, but in general still represents a second integral of the equivalent, extended, time-independent Hamiltonian system, in which ψ is regarded as the Hamiltonian function. Hamiltonian theory would thus appear to indicate that the existence of a materially conserved quantity should render the flow associated with a given streamfunction integrable, as described above in Chapter 3. This in turn would appear to suggest that the trajectories in inviscid, two-dimensional turbulence could not, in a strict sense, be chaotic – that is, could not include a chaotic invariant set— except in regions in which the potential vorticity is uniform in space and time. As anticipated by the discussion in Section 3.7, the correct conclusion is that vorticity-conserving, two-dimensional flow is always integrable in the following specific sense: if the vorticity q and streamfunction ψ are both known as functions of space (x, y) and time t, then the trajectories of the fluid motion can be obtained directly from these two functions in any region of nonvanishing vorticity gradient, without integrating the differential equations that relate the trajectories to the velocity field (Brown

& Samelson (1994)). It is important to emphasize that, if the time domain is not compact, integrability in this sense does not imply that the motions are periodic or quasiperiodic, and does not rule out the possibility that trajectories in two-dimensional turbulence may be, in the strict sense, chaotic. Balasuriya [2001] has considered some of these issues further. Note that, if chaotic trajectories did exist in a region of nonvanishing vorticity gradients, it could be inferred that the vorticity field would become discontinuous in the limit $t \rightarrow \infty$. Since the vorticity field could still be smooth for any finite time t, this would not be in conflict with existing global regularity results for two-dimensional, incompressible flow.

This surprising result that, in a carefully defined sense, a two-dimensional, turbulent flow may be simultaneously both chaotic and integrable, resolves the apparent paradox raised by the vorticity conservation equation (7.1) and the Hamiltonian analogy. It also serves to illustrate that while results obtained from the Lagrangian perspective must always be consistent with existing Eulerian results, they can nonetheless offer unexpected new insights into the properties of geophysical fluid flow. Recent research, including the articles summarized in the following sections, confirms that the fundamental structures and mechanisms of Lagrangian transport found in kinematic models also arise in numerical solutions of geophysical fluid-dynamical models that explicitly include the nonlinear dynamical advection of potential vorticity by the Lagrangian flow, and in related laboratory flows.

7.2 Barotropic Jets

The qualitative picture of fluid motion in meandering jets and traveling waves built up in the previous chapters was based entirely on kinematic models. It should not be surprising that the results of both analytical and numerical studies of subsequent barotropic dynamical models of traveling waves and meandering jets have tended to confirm the qualitative picture provided by the kinematic models, provided that the dynamical models are considered in regimes in which the corresponding dynamical streamfunctions qualitatively resemble the kinematic streamfunctions. Two early studies that provided this confirmation considered Rossby-wave critical layerscritical layer (Ngan & Shepherd (1997)) and numerical simulations of a fully nonlinear, unstable jet (Rogerson et al. (1999), Miller et al. (1997); see also Pierrehumbert (1991b)).

The numerical solutions for the unstable barotropic jet were obtained in a parameter regime in which the instabilities rapidly saturate and the flow enters a near-periodic, oscillatory phase (Figure 7.1), with a meandering jet that is qualitatively and quantitatively similar to the kinematic model (1.27). Miller et al. (1997) and Rogerson et al. (1999) developed numerical reconstructions of hyperbolic trajectories and their associated material

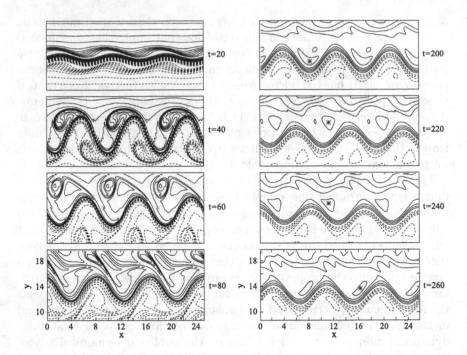

FIGURE 7.1. Streamfunction contours for the quasigeostrophic barotropic jet during the evolution toward a near-periodic oscillatory phase. From Rogerson et al. (1999).

manifolds in the near-periodic oscillatory phase, and used lobe transport to compute transports associated with these manifolds (Figure 7.2).

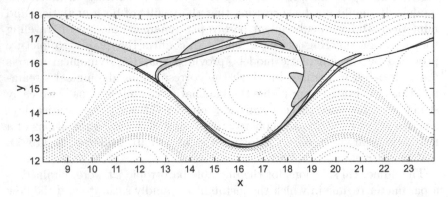

FIGURE 7.2. Streamfunction contours (dotted lines), material manifolds (thick solid), and lobes (shaded regions) for the quasigeostrophic barotropic jet in the near-periodic oscillatory phase. After Rogerson et al. (1999).

This analysis, the first of its kind for a numerically simulated geophysical fluid flow, revealed Lagrangian transport patterns similar to those obtained from the kinematic model (1.27). The transport occurred primarily along the edges of the jet, around the recirculation regimes induced by the propagating meanders. Exchange between the jet and the recirculation cells was weaker. For dimensional parameters representative of the Gulf Stream, lobe-transport estimates of the exchange between the external, retrograde regime and the recirculation cells were estimated at 4 to 5×10^6 $m^3\, s^{-1}$. Rogerson et al. (1999) noted that these values were similar, for example, to the fluxes associated with Gulf Stream rings, which are well-known, long-lived coherent eddy structures that result from the pinch-off of nonlinear meanders and are believed to dominate the related, but qualitatively different, transport of near-surface fluid from one side of the stream to the other.

These barotropic calculations, and the associated stable and unstable manifolds of hyperbolic points in a comoving frame, and the corresponding lobes associated with manifold intersections, have proven useful also for the interpretation of entrainment and detrainment events in observed Gulf Stream float trajectories. Lozier et al. (1997) used data from the 37 floats launched as part of the RAFOS pilot program (Bower et al. (1986)) conducted over the years 1984–1985. Phase speed information during the time of the float deployment was obtained from work of Lee (1994) and Lee & Cornillon (1995), who used satellite AVHRR (advanced very high resolution radiometer) infrared images of the sea surface temperature to digitize the Gulf Stream path for the period April 1982 to December 1989. At any given time the Gulf Stream is characterized by a number of different meanders that move at different phase speeds. Lozier et al. (1997) developed a technique for obtaining a spatial and temporal match between a specific meander and a given float's position, subject to a variety of criteria, which ultimately were satisfied by only 4 of the 37 floats. For these 4 floats, it was possible to transform the trajectory to a frame moving with the appropriate meander phase velocity.

In the moving frames, looping and oscillatory behavior was observed that was strongly suggestive of motion in and out of finite-amplitude recirculation structures similar to those seen in the crests and troughs of the meanders of the kinematic and dynamical barotropic jet models. This behavior was further correlated with the stable and unstable manifolds in the numerical model of a jet described in Miller et al. (1997), and it was concluded that crossings of critical lines (i.e., regions where the meander phase speed equals the maximum zonal velocity of the jet) are associated with lobes of fluid that move in and out if the recirculations. During the transit times of these floats, there were no observed ring formations or ring-stream interactions, so the velocity field should have been relatively well represented, at least qualitatively and perhaps even quantitatively, by both the kinematic and the dynamical jet models described above. Lozier et al.

(1997) also point out that the geometrical features of the lobes are reminiscent of the filaments that have been observed trailing from Gulf Stream meander crests and troughs in satellite imagery of sea-surface temperature. This work suggested that the deformation of lobes that results from their interaction with hyperbolic trajectories should enhance the diffusion of any properties carried by the fluid mass contained in the lobes.

The innovative use of a comoving frame of reference in the study by Lozier et al. (1997) was motivated by the development of the dynamical systems approach to transport in jet models that is reviewed in this text. It indicated that, in an appropriate moving frame, there are geometrical structures in the flow that can be used to analyze and interpret oceanographic observations of Lagrangian motion. More generally, it also suggests that there are situations in which transport phenomena induced by the large-scale geometrical structures in the oceanic flow field are not dominated by smaller-scale turbulent mixing. At the same time, it suggests that the efficiency and effects of the small-scale mixing may be influenced and even partially controlled by the large-scale Lagrangian motion.

In the Rossby-wave critical-layer study (Ngan & Shepherd (1997)), the consistency of the kinematic and dynamical models was an explicit result of the analytical theory. For an incident, neutral Rossby wave in a mean zonal (x-directed) shear flow $U(y)$, a critical layer arises at any point $y = y_c$ where the phase speed c of the zonally propagating neutral wave matches the mean-flow zonal velocity $U(y_c)$. At that point, since $U - c$ vanishes, the linearized potential vorticity equation (7.1) is singular. As shown by Stewartson (1981) and Warn & Warn (1978), a matched-asymptotic analysis can be conducted around $y = y_c$ to determine the evolution of the nonlinear solution. The streamfunction in and around the critical layer has a form similar to that of the recirculation cell and surrounding shear flow in the kinematic model (1.1) of the meandering jet, with separatrices connecting hyperbolic points and dividing the domain into an interior recirculation regime and two external zonal flow regimes. Remarkably, the separation of scales between the streamfunction and the vorticity near the critical layer means that, to first order in the expansion parameter, the streamfunction is dominated by the larger-scale shear flow and is not affected by the redistribution of vorticity within the critical-layer region. This provides an explicit example of a dynamical model in which the potential vorticity field is advected kinematically, that is, it is advected by a velocity field with a known structure that is itself independent of the resulting potential vorticity advection.

In their analysis, Ngan & Shepherd (1997) used lobe transport to analyze the Lagrangian motion that results when the Stewartson–Warn–Warn critical-layer solution is perturbed by an additional, weak, externally forced Rossby wave. As anticipated, this additional forcing causes the breakup of the separatrices and the development of patterns of Lagrangian motion similar to those in the kinematic model. Note that, in the critical-layer case, the recirculation zones form asymptotically narrow "cat's eyes" around $y = y_c$, rather than the finite-amplitude structures in the kinematic and the barotropic numerical models discussed above. The critical-layer study, which was primarily motivated by the problem of the mixing of ozone and other chemical species across the stratospheric polar jet, also addressed some aspects of the transient mixing of a passive tracer (which in this case also includes vorticity). This mixing was observed to occur in two stages, with a rapid initial cascade of variance to small spatial scales, followed by homogenization at the small scales. Ngan & Shepherd (1997) note that this behavior is clearly different from diffusive mixing, which occurs first at the smallest scales. This study thus again illustrates that the efficiency and effects of small-scale mixing may be influenced or controlled by the large-scale Lagrangian motion.

A number of other studes have used related geometric methods to study Lagrangian processes in barotropic models of the atmosphere. Koh & Plumb (2000) applied lobe transport to a contour-dynamical, barotropic, numerical model of stratospheric Rossby-wave breaking. Pierrehumbert (1991b) used Lyapunov exponents to diagnose chaotic advection in the traveling wave example, with wave parameters derived from linear quasigeostrophic theory. Bowman (1993) conducted numerical simulations for an idealized, two-dimensional model of the stratospheric polar vortex, and studied the kinematics and dynamics of the fluid exchange caused by interactions of large-scale, propagating waves with the jet.

Mixing and stirring in barotropic jets has also been studied in the laboratory. Sommeria et al. (1989) conducted experiments in a rotating annulus, and obtained remarkable results that strikingly demonstrated the ability of a meandering jet to mix properties along its flanks while maintaining sharp property gradients across its core. These experiments motivated del Castillo-Negrete & Morrison (1993) to study transport by traveling waves in a mean shear flow, using methods adapted from Hamiltonian mechanics and model wave parameters derived from linear quasigeostrophic theory.

7.3   Baroclinic Jets

The barotropic models discussed above provide a natural starting point for the exploration of the Lagrangian transport mechanisms in dynamical models. However, the Gulf Stream and many other geophysical jets are known to have three-dimensional spatial structure that cannot be represented in the two-dimensional barotropic models. Observations (e.g., Bower et al. (1986)) show that the characteristics of Lagrangian motion in the Gulf Stream also change substantially with depth, with more exchange occurring at deeper levels, and less near the surface. Associated with this vertical differentiation in the character of the lateral Lagrangian motions is a concurrent depth-dependence of the observed mean cross-stream potential vorticity gradient, with the deeper gradients being weak, and the shallow gradients strong (Bower et al. (1986)). These observations raise a question of cause and effect: does the strong potential vorticity gradient act as a "barrier" that prevents shallow cross-stream motion, or does the tendency for cross-stream motion at depth erase the deep potential vorticity gradient? A related question concerns the potential role of dynamical baroclinic instability processes, which arise from vertical gradients in the mean horizontal velocity, and so cannot occur in barotropic models, but are known to play a significant role in the development and evolution of Gulf Stream meanders (e.g., Cronin & Watts (1996)).

Consideration of the Lagrangian motion in kinematic models such as (1.27) suggests one possible answer to these questions (Bower (1991), Samelson (1992), del Castillo-Negrete & Morrison (1993)). The Lagrangian analysis clearly identifies the material manifolds surrounding the recirculation regimes as the sites at which fluid exchange occurs in the kinematic models. Thus, the geometric structure of these material manifolds in the unperturbed flows strongly affects the character of the mixing that occurs in the perturbed flows. Cross-jet exchange is relatively weak when the recirculation regimes on opposite sides of the jet are far apart and isolated from one another, and relatively strong when they are close or even adjacent. In the extreme case of separatrix reconnection, in which the recirculation regimes on opposite sides of the jet come into direct contact, and the central jet regime disappears, exchange across the entire jet is strongly enhanced. Since the locations of the recirculation cells are controlled by the critical-layer condition $U(y_c) = c$, the strength of cross-stream exchange in the perturbed flow should increase as the phase speed of the basic meander or traveling wave increases toward the maximum speed of the jet. In the case of a baroclinic current such as the Gulf Stream, the depth-dependence of the mean flow velocity will result in a corresponding depth-dependence of the relative velocity of the phase speed of a traveling wave or meander, even though the absolute phase speed of the coherent wave disturbance is independent of depth. If the mean current decreases with increasing depth, as it does in the Gulf Stream, the recirculation regimes induced by the traveling disturbance should approach each other with increasing depth. Thus, the models suggest that cross-stream exchange should increase with depth purely as a kinematic consequence of the depth variation of the relative phase speed of the traveling wave and the maximum mean current (Bower (1991), Samelson (1992), del Castillo-Negrete & Morrison (1993)). This prediction of enhanced cross-jet exchange with depth is consistent with existing observations (Bower et al. (1986)).

This qualitative inference from the kinematic models has been supported by results from several different dynamical models. Numerical simulations with a barotropic model, in which changes in relative phase speed of the jet and traveling disturbance are achieved by varying the ambient potential vorticity gradient, show a similar pattern, with strong cross-stream exchange when the wave phase speed is close to the jet speed, and weak cross-stream exchange when the wave phase speed is far from the jet speed (Yuan et al. (2001)). These barotropic simulations, however, cannot directly address the possible role of baroclinic instability. Linearized dynamical analysis, for a mean flow with vertical but no horizontal shear, suggests that even in the presence of baroclinic instability, it is the analogous critical-level condition $U(z_c) = c$ (for flows with mean velocity U depending on depth z) that determines the level at which cross-stream motion is maximum, and not the level of minimum cross-stream potential vorticity gradient (Lozier & Bercovici (1992)). Numerical simulations of two-layer quasigeostrophic flows with both vertical and horizontal shear in the mean horizontal jet velocity have recently been performed that provide the first confirmation of these ideas in a nonlinear, baroclinically unstable model (Yuan et al. (2004)). The analysis of material manifolds and particle motions in these simulations suggests that a transition from weak to strong cross-stream motion in the lower layer occurs for mean-flow parameters that are representative of the observed vertical structure of the Gulf Stream (Yuan et al. (2004)). The transition is marked by the onset of baroclinic instability, which in turn drives cross-stream motion in the lower layer (Yuan et al. (2004)). In the upper layer, the material manifolds associated with the recirculation cells on opposite sides of the stream remain isolated from one another, and the cross-stream potential vorticity gradient remains strong; in the lower layer, the corresponding material manifolds become entangled, and the cross-stream potential vorticity gradients are essentially erased.

Related atmospheric studies of mixing in baroclinic flows exist. For example, the nonlinear analytical solution of a weakly nonlinear two-layer flow by Warn & Gauthier (1989) displays some characteristics that have a qualitative similarity to the baroclinic jet models discussed above. Analyses using observationally based representations of atmospheric wind fields have also been carried out: for example, Joseph & Legras (2002) and Bowman (2000) used Lyapunov-exponent and related methods to estimate material manifolds and associated atmospheric transport from wind fields estimated using operational numerical weather prediction models.

7.4   Boundary Currents and Recirculations

The development and discussion of the preceding chapters and sections have focused on Lagrangian motion in traveling waves and meanders in zonal jets. When one side of a jet is confined by a rigid boundary, the flow generally loses its cross-jet symmetry, and the character of the time-dependent motion can be quite different. Nonetheless, numerical and laboratory studies show that the present perspective still gives insight into the Lagrangian transport processes in these partially confined jets. Two examples of this type of flow are briefly summarized here.

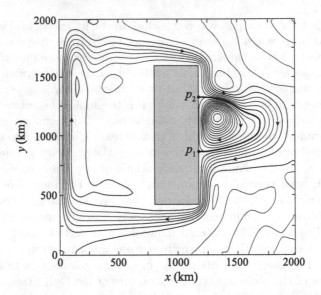

FIGURE 7.3. Streamlines for the steady large-scale wind-forced flow with the nonlinearity parameter $r_\delta = 1.0$. A separatrix connects the hyperbolic stagnation points p_1 and p_2, defining the recirculation region in the steady flow. For larger values of the parameter r_δ, the flow becomes unsteady, resulting in exchange of fluid between the recirculation and the surrounding region. From Miller et al. (2002).

The first example, from Miller et al. (2002), involves the transport associated with a recirculation region along the eastern boundary of a meridionally elongated island in a numerical model of large-scale wind-forced flow in an ocean basin. The character of the flow depends on the model nonlinearity parameter r_δ. For the value $r_\delta = 1.0$, the flow is steady, and a separatrix connects two hyperbolic stagnation points along the boundary, defining a recirculation region in the steady flow (Figure 7.3). For larger values of the parameter r_δ, the flow becomes unsteady, resulting in exchange of fluid between the recirculation and the surrounding region. Miller et al. (2002) showed that hyperbolic trajectories could be identified in the unsteady flow, and that the exchange between the recirculation and the surrounding region could be quantified using lobe transport analysis (Figure 7.4). For sufficiently large values of the nonlinear parameter r_δ, the lobe transport dominates the Ekman transport (Figure 7.5). Miller et al. (2002) also show that potential vorticity is advected into and out of the gyre by lobes, and give methods to calculate these fluxes explicitly. These fluxes play an important dynamical role, by changing the circulation about the gyre.

Miller et al. (2002) also conducted related laboratory experiments, including dye visualizations of the unstable manifold of the upper hyperbolic

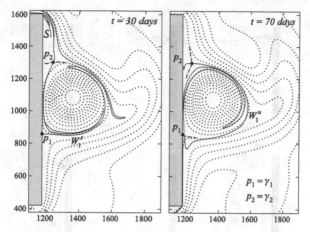

FIGURE 7.4. Streamfunction contours (dotted lines) and lobe geometry (solid) at times $t = 30$ days and $t = 70$ days for the large-scale wind-driven flow with nonlinearity parameter r_δ larger than the instability threshold, for which the flow becomes time-dependent. From Miller et al. (2002).

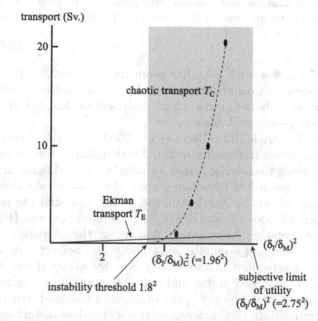

FIGURE 7.5. Fluid exchange due to lobe transport (T_C, labeled "Chaotic Transport" by Miller et al. (2002)) and Ekman transport (T_E) in units of Sv (10^6 m^3 s^{-1}) as a function of the square of the nonlinearity parameter $r_\delta = \delta_I/\delta_M$, which measures the ratio of inertial and viscous boundary layer scales. The shaded region indicates the region of r_δ where a meaningful recirculation zone could be defined using stable and unstable manifolds of hyperbolic trajectories. From Miller et al. (2002).

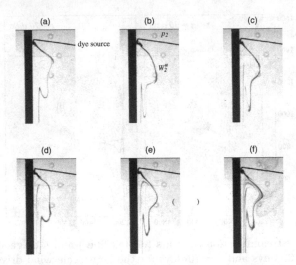

FIGURE 7.6. A sequence of video images (time increases from (a) to (f)) showing the unstable manifold, visualized with dye, of the upper hyperbolic trajectory. Note the characteristic looping structures associated with lobe formation. Additional flow visualizations and analysis were performed to locate the hyperbolic trajectory and to ensure that the dye source was placed correctly. From Miller et al. (2002).

trajectory for a flow with a similar geometry (Figure 7.6). These experiments provide direct confirmation that the present methods and concepts are relevant not only to numerical and analytical models, but also to physical models of geophysical flows.

The second example, from Deese et al. (2002), uses laboratory methods to study Lagrangian transport associated with multiple, time-dependent recirculations along the interior side of a similarly confined boundary current. This work was motivated by oceanographic observations of a deep western boundary current in the North Atlantic, which indicated the presence of multiple recirculations offshore of the deep boundary current (Figure 7.7). These counterrotating recirculation regions on the offshore edge of the boundary current had strengths of 4 and 8 Sv, respectively, and must apparently be separated in the steady flow by a hyperbolic stagnation point and its associated stable and unstable manifolds (Figure 7.8a). If only one such recirculation were present, no such manifold structure would exist (Figure 7.8b). In the former case, time-dependent disturbances to the steady flow can be expected to induce exchange between the two recirculation regions and between the recirculations and the boundary current (Figure 7.8c). In the latter case, with only one recirculation, the absence of hyperbolic points and the associated manifold structure makes the Lagrangian motion less sensitive to time-dependent disturbances, and it can be anticipated that regime exchange will be more difficult to induce.

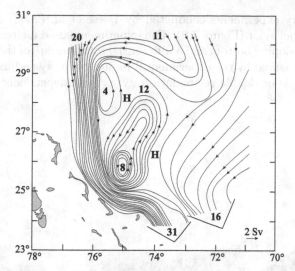

FIGURE 7.7. Observational estimates of the streamlines of the time-averaged flow at 1000-m depth near 25° N, 75° W in the North Atlantic, associated with the deep western boundary current in the vicinity of Abaco Island. The contour interval is 2 Sv ($1\,\mathrm{Sv} = 10^6\,\mathrm{m}^3\,\mathrm{s}^{-1}$). Possible locations of saddle-type (hyperbolic) stagnation points are indicated by an H. From Deese et al. (2002).

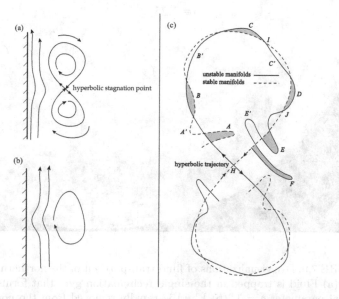

FIGURE 7.8. Schematic illustrations of the steady streamline geometry associated with (a) double and (b) single recirculation gyres, respectively, adjacent to a western boundary current. (c) Unsteadiness on the twin recirculation gyre causes breakup of the regime boundaries, and fluid exchange through lobe transport by the lobes labeled $A, A', B, B', C, C', D, D', E, E', F, \ldots$. From Deese et al. (2002).

Laboratory experiments conducted by Deese et al. (2002) confirm this qualitative behavior (Figure 7.9). The experiments used a circular rotating tank with a sloping bottom and a differential rotation rate of the lid. The lid rotation, measured by the experimental parameter δ, was adjusted to force either a single or double recirculation gyre. Time-dependence of the flow

$\delta = 1.00$

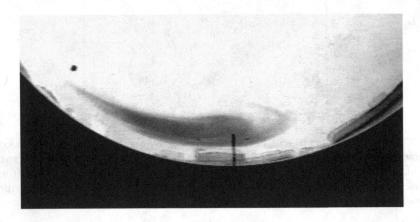

$\delta = 1.25$

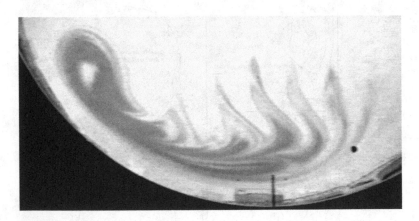

FIGURE 7.9. Dye visualizations of fluid transport out of the northern (left-hand) gyre. (a) Fluid is trapped in the single recirculation gyre that forms for experimental parameter $\delta = 1$. (b) Fluid is rapidly removed from the northern ends of the two recirculation gyres that form for $\delta = 1.25$, which the characteristic looping structures of lobe transport. Both flows are unsteady (time-periodic), and north is to the left. From Deese et al. (2002).

was introduced by introducing an additional modulation of the lid rotation. For $\delta = 1$, the system has a single recirculation, and dye visualizations indicate that time-dependent modulation does not efficiently remove fluid from that recirculation (Figure 7.9a). For $\delta = 1.25$, the system has a double recirculation, and dye visualizations show relatively rapid removal of fluid from the northern recirculation, with the characteristic looping structures of lobe transport (Figure 7.9b).

For the latter case, Deese et al. (2002) were able to track individual lobes approximately, and to verify that the lobe transport followed the qualitative pattern indicated in the schematic illustration (Figure 7.8c). They also showed that the exchange between the western boundary layer and the recirculation gyres occurred by lobe transport.

A number of studies have begun to explore the use of manifold-based methods in the simulated western boundary current extensions in idealized basin models. These exploratory studies have served as a useful test-bed for numerical techniques, and have provided some suggestive physical results. Poje & Haller (1999) used manifold-based methods to analyze ring detachment and merger events in a double-gyre ocean model based on the shallow-water equations. This model included a time-dependent layer thickness, so the two-component horizontal velocity field was not strictly incompressible. Coulliette & Wiggins (2001) studied intergyre exchange in a three-layer, quasigeostrophic, double-gyre ocean model, for velocity fields with periodic, quasiperiodic, and aperiodic time-dependence, and compared the effectiveness of the manifold-based methods in a sequence of simulations with increasing spatiotemporal complexity.

7.5   Advanced Topics

The methods described in this text have only recently begun to be used for the analysis of geophysical fluid flows. In this section, brief discussions of a number of general topics relating to current or future research questions are collected. These involve both technical and conceptual issues.

Discrete Finite-Time Velocity Fields

The mathematical methods of dynamical systems theory were originally developed for the analysis of velocity fields that are defined continuously and for all time over the entire fluid domain. Largely because of this history, initial efforts to use dynamical systems theory to analyze Lagrangian transport in geophysical fluid models proceeded within the mathematically convenient but restrictive framework of time-periodic flows with velocity fields defined by analytic functions. In general, however, only highly idealized physical models give rise to velocity fields of this type. Flows that are generated by numerical models or observed by geophysical measurements will always be defined by a finite set of numbers, for a finite interval of time, on discrete spatial and temporal grids.

The differences between discrete and continuous spatial or temporal representations are relatively easy to overcome. Any of a variety of standard interpolation, smoothing, and integration techniques, may be used to extend the discrete record to a consistent continuous analogue, and to solve the appropriate equations for a consistent set of fluid trajectories. In any given case, of course, the accuracy of the results will depend on the characteristics of the techniques that are used, as well as the accuracy and resolution of the original numerical or observational record. Some relevant numerical techniques are discussed in Appendix C.

The restriction to finite time intervals is potentially more fundamental. Many of the mathematical objects of interest in dynamical systems theory are asymptotic in nature, and so require infinite time intervals for their very definition. Consequently, these objects, and the possibility that they may provide useful insights into the Lagrangian fluid motion, threaten to fade into nothingness in the face of the ephemeral nature of any geophysical fluid model or flow. What saves them is what saves all such asymptotic ideas in mathematical physics: when the time scales of the finite-time physical flow are—in some appropriate sense—sufficiently long, the asymptotic concepts and results provide both fundamental qualitative insights into flow characteristics, and useful quantitative measures of flow properties. It is also possible to extend the mathematical theory to treat finite-time intervals explicitly. For example, recent numerical and theoretical approaches to the definition and identification of hyperbolic trajectories and their stable and unstable manifolds in finite-time velocity fields include Haller (2000), Haller (2001), Haller & Yuan (2000), Haller (2002), Ide et al. (2002), Mancho et al. (2003), Mancho et al. (2004), and Ju et al. (2003).

In general, many theorems of dynamical systems theory have a geometrical character, and do not require explicit analytic forms for the velocity fields. For example, the existence of the associated stable and unstable manifolds follows directly from the existence of a hyperbolic trajectory, regardless of the specific structure of the trajectory and manifolds. Such results provide general conceptual guidance that can be combined with appropriate numerical techniques to develop geometrically based analysis methods for velocity fields defined entirely by discrete numerical representations. These methods can be used for the analysis of Lagrangian transport in any real or simulated geophysical flow for which the velocity field can be adequately quantified, either from numerical model output or by a sufficiently dense and accurate set of measurements.

Boundary Conditions

For many flows, some part of the domain may have rigid boundaries. The corresponding boundary conditions imposed on the velocity field are most frequently either free-slip, in which the velocity normal to the boundary is required to vanish, or no-slip, in which the full velocity is required to vanish. For free-slip boundary conditions, the dynamical systems methods described in this text are generally appropriate even at the rigid boundary, since under these conditions the rigid boundary itself naturally forms an invariant material manifold. Regions of separation of fluid from the boundary or convergence of fluid toward the boundary may indicate the presence of hyperbolic trajectories on the boundary.

For no-slip boundary conditions, different technical issues arise. In this case, an analogue of saddle-points often appears on the boundary, but these cannot be hyperbolic, because the velocity along the boundary is identically zero. Some theorems on existence of stable and unstable material manifolds of these nonhyperbolic points exist that may be applicable to two-dimensional time-periodic velocity fields with no-slip boundary conditions (McGehee (1973), Casasayas et al. (1992) , Fontich (1999) and Yuster & Hackborn (1997)). These manifolds are also relevant to transport analysis: they are material curves and can intersect to form lobes. However, their asymptotic properties are different: points in the stable and unstable manifolds approach the nonhyperbolic point at algebraic, rather than exponential, rates as $t \to \infty$ and $t \to -\infty$, respectively.

Three-Dimensional Flows

Many important geophysical flows are intrinsically three-dimensional. It is straightforward to extend the definitions of hyperbolic trajectories and stable and unstable material manifolds to three-dimensional flows. In three dimensions, however, the sum of the dimensions of the stable and unstable manifolds of a hyperbolic stagnation point must itself be three. For incompressible flows, for which a hyperbolic stagnation point must always have both a stable and an unstable manifold, this means that one of the two manifolds must be one-dimensional. Since a one-dimensional material manifold is not a flow barrier in three dimensions, this difference is fundamental. For essentially the same reason, chaotic trajectories are possible in three-dimensional, but not two-dimensional, steady flow.

A related difference between two- and three-dimensional flows is that the analogy between incompressible flow and classical Hamiltonian dynamics (Section 3.7) is lost in three dimensions. Under certain conditions, a coordinate system can be found in which some of the features of the Hamiltonian analogy are retained, as shown by Mezic & Wiggins (1994) and Haller & Mezic (1998). Also, a KAM theorem for time-quasiperiodic flows does exist for three-dimensional flows (Broer et al. (1996)). However, there are many more possibilities for material surfaces and for more exotic types of lobe dynamics, as well as a variety of new mechanisms for chaos. Dynamical systems methods have been used to study transport in three dimensions by Cartwright et al. (1994, 1995, 1996), Mezic et al. (1998), Yannacopoulos et al. (1998), Fountain et al. (2000), Haller (2001), and Mezic (2001a,b).

Unresolved Scales of Motion

Geophysical fluid motions generally possess a large and continuous range of temporal and spatial scales. Models of these phenomena, such as the meandering jet or the traveling wave flows considered here, involve a variety of approximations and simplifications. It is necessary to keep in mind these approximations and simplifications, and the limitations of the resulting models, in relating the results of such analyses to the original physical situations.

These considerations are particularly important when the velocity field of interest is derived from a dynamical model that includes a specific representation or "parametrization" of the aggregate effects of unresolved scales of motion, such as an eddy viscosity. In that case, the parametrized fluxes of vorticity or other dynamical quantities may be as large as, or even larger than, the advective fluxes by the resolved motions. If the parametrized fluxes are supposed to represent advective fluxes on the unresolved scales, then the meaning of the computed trajectories is in doubt: most of the true physical motion must apparently occur at smaller scales. An extreme example of such a circumstance would be a computation of trajectories in a linearized ocean gyre model such as the Munk (1950) model, in which the vorticity balance in the western boundary layer is dominated by meridional motion in the ambient gradient, and viscous diffusion. In that case, the viscous diffusion is assumed to be many orders of magnitude larger than molecular diffusion, and its physical interpretation is not clear; often, the value of the diffusion coefficient is set simply by the requirement that the resulting boundary layer have approximately the width of an observed western boundary current such as the Gulf Stream, rather than by any meaningful understanding of the smaller-scale physical processes that are represented by the diffusive terms.

Unfortunately, the physical situation is in general sufficiently complex that the analysis and representation often cannot be unambiguously tested simply by refining the resolution of a given numerical simulation, such as that of the Munk (1950) model, with the hope of explicitly computing the parametrized fluxes. Such a refinement may not even lead to a physically reasonable solution; see, for example, the discussion in Pedlosky (1996). In general, the range of scales is so great in geophysical flows that the parametrized physical processes may be quite different in character than the resolved processes; for example, they may effectively include three-dimensional effects that have no explicit counterpart in a two-dimensional dynamical model. This complexity limits the physical insight that can be achieved, for example, through direct tests of convergence by grid refinement. As is often the case with geophysical fluid models, careful thought and physical interpretation of the results is necessary and important. The presumption of this text is that, in a broad sense, the present methods should prove useful for the analysis of any flow that contains coherent structures, such as traveling waves or meandering jets, and in which advective processes on the scales of those coherent features are important to the dynamics and transport mechanisms of the flows.

7.6 Summary

The material in this chapter necessarily represents only a brief overview of a few examples from the evolving range of studies in geophysical fluid dynamics that make use of the conceptual approach outlined in this text. It is clear already from these results that new insights can be gained into the kinematics and dynamics of jets, waves, and other geophysical flows using this approach. It is equally clear that the concepts and numerical techniques that allow systematic investigations of this type are only now emerging. In a general way, this perspective serves also to illustrate and emphasize that the relations between Eulerian and Lagrangian motions, and between their different representations in numerical simulations of geophysical flows, can be subtle and still offer much room for exploration. We believe that this approach to the analysis of fluid flows promises many more insights into the motions that surround us in the ocean and atmosphere, and we hope that the material presented here will be useful to those who seek those insights.

Appendix A
Mathematical Properties of Fluid Trajectories

Existence, Uniqueness, and Differentiability

Consider the trajectory equations

$$\dot{\mathbf{x}} = \mathbf{v}(\mathbf{x}, t), \qquad (A.1)$$

where the velocity field $\mathbf{v}(\mathbf{x}, t)$ is C^r, $r \geq 1$, on some open set U in $(n + 1)$-dimensional space, $\mathbf{x}$ and $\mathbf{v}$ are n-dimensional vectors, and the scalar variable t represents time.[1] Here n can be either two or three, and U is the space-time domain for the flow. Let $\mathbf{x}(t; t_0, \mathbf{a})$ denote the trajectory through the point $\mathbf{a}$ at $t = t_0$.

From the point of view of dynamical systems theory, (A.1) is a set of ordinary differential equations for the fluid trajectories. Proofs of the following fundamental statements can be found in many ordinary differential equations textbooks, for example, Arnold (1973), Hirsch & Smale (1974), or Hale (1980).

- Given any point $\mathbf{a}$, there exists a trajectory $\mathbf{x}(t; t_0, \mathbf{a})$ passing through this point at $t = t_0$.

- The trajectory $\mathbf{x}(t; t_0, \mathbf{a})$ exists in time t until $\mathbf{x}$ either leaves the domain, or becomes infinite.

- Trajectories are always unique in the sense that there is only one trajectory that passes through a given point at a specific time. However, if the velocity field is steady, then all trajectories passing through a given point $\mathbf{a}$ are equivalent, except for their differing offsets in time. If the velocity field is unsteady, the trajectories passing through $\mathbf{a}$ typically are not equivalent in this sense.

- Trajectories are C^r functions of t, t_0, and $\mathbf{a}$.

---

[1] C^r means that $\mathbf{v}(\mathbf{x}, t)$ can be differentiated r times with respect to $\mathbf{x}$ and t, and that each of these derivatives varies continuously with respect to $\mathbf{x}$ and t.

Volume Preservation

Let D_{t_0} denote a domain in n-dimensional space at time t_0, and let D_t denote the evolution of D_{t_0} under the transformation that maps points along fluid trajectories $\mathbf{x}(t, t_0, \mathbf{a})$. Let $V(t)$ denote the volume of D_t. Then

$$\frac{dV}{dt}\bigg|_{t=t_0} = \int_{D_{t_0}} \nabla \cdot \mathbf{v} \, d\mathbf{x}, \tag{A.2}$$

where $\nabla \cdot \mathbf{v} = \partial v_1/\partial x_1 + \cdots + \partial v_n/\partial x_n$ denotes the divergence of the n-dimensional velocity field.

Suppose that the divergence is everywhere constant, i.e.,

$$\nabla \cdot \mathbf{v} = C = \text{constant.} \tag{A.3}$$

Since t_0 is arbitrary, the evolution equation for the volume can then be obtained from (A.2). This yields

$$\dot{V} = CV,$$

which has the obvious solution

$$V(t) = e^{Ct}V(0). \tag{A.4}$$

If the flow is incompressible (i.e., $C = 0$), it follows that

$$V(t) = V(0).$$

Thus, incompressible flows preserve volume.

One can also characterize volume preservation in terms of the Jacobian of the transformation that maps points along particle trajectories $\mathbf{x}(t, t_0, \mathbf{a})$. Using the standard formula from calculus for the change of volumes under coordinate transformations, we have

$$V(t) = \int_{V(t)} d\mathbf{x} = \int_{V(t_0)} \det\left(\frac{\partial \mathbf{x}}{\partial \mathbf{a}}(t, t_0, \mathbf{a})\right) d\mathbf{a} = V(t_0). \tag{A.5}$$

Hence if $\det\left(\frac{\partial \mathbf{x}}{\partial \mathbf{a}}(t, t_0, \mathbf{a})\right) = 1$, the volume of a region of the fluid is unchanged as it evolves in the flow.

Behavior of Fluid Trajectories Under Coordinate Transformations

A judicious choice of coordinate transformation is often helpful, as in the example (4.5) in Chapter 4. In Ide et al. (2002), it is shown that (1) trajectories always transform to trajectories, even under nonlinear and time-dependent coordinate transformations; and (2) hyperbolic trajectories transform to hyperbolic trajectories, provided that the coordinate transformation does not depend exponentially on time.

Appendix B

Action-Angle Coordinates

In this appendix, some additional details of the transformation to action-angle coordinates are provided, and an example is explicitly computed. The traditional development of action-angle coordinates utilizes the idea of generating functions from classical mechanics (see, e.g., Goldstein (1980) or Landau & Lifschitz (1976)). An excellent classical derivation of action-angle coordinates is given by Percival & Richards (1982). The approach taken here (and loosely inspired by Melnikov (1963); see Wiggins (2003) for details) more directly utilizes the underlying geometry of the streamlines of the velocity field.

Action-angle coordinates can be constructed in regions of a two-dimensional steady, incompressible flow containing closed streamlines. Recall that for the steady traveling wave flow in the comoving frame, there were two types of regions of closed streamlines: the recirculation regions and the jet itself; the jet could be considered a region of closed streamlines by virtue of its spatial periodicity. These regions of closed streamlines were separated by heteroclinic trajectories.

Here we define the transformation to the action and the angle coordinates, and describe their relationship to the geometry of closed streamlines. We consider a region of closed streamlines surrounding an elliptic stagnation point, (x_c, y_c). Let L denote a reference curve that emanates from (x_c, y_c) and intersects each closed streamline (periodic orbit) only once; see Figure B.1.

We denote L in parametric form by $[x_0(s), y_0(s)]$, $s_1 \leq s \leq s_2$, where s_1 and s_2 are the endpoints of the parameter interval, and solutions of

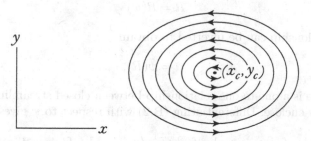

FIGURE B.1. The reference curve L with respect to which the angle variable associated with a given closed trajectory is defined.

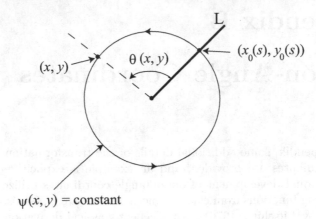

$$\psi(x, y) = \text{constant}$$

FIGURE B.2. Definition of the angle variable.

(3.26) starting on L by $(x(t, s), y(t, s))$ where $x(0, s) = x_0(s)$ and $y(0, s) = y_0(s)$. Let (x, y) be a point on a closed streamline of $(x(t, s), y(t, s))$ and let $t = t(x, y)$ be the time taken for the trajectory starting at $(x_0(s), y_0(s))$ to flow to (x, y). We denote the period of each periodic orbit defined by $\psi(x, y) = \psi = $ constant by $T(\psi)$. The *angle* variable, $\theta(x, y)$, is now defined as

$$\theta(x, y) = \frac{2\pi}{T(\psi)} t(x, y), \tag{B.1}$$

where $(x, y) \in \psi = $ constant; see Figure B.2.

The action variable $R(x, y)$ is simply the area enclosed by any closed streamline,

$$R = \frac{1}{2\pi} \oint_{\psi} x \, dy, \tag{B.2}$$

where ψ labels the closed streamline defined by $\psi(x, y) = \psi = $ constant. Clearly, since ψ is time-independent, R is also time-independent. From (B.2), the action variable for a closed streamline is a function of ψ,

$$R = R(\psi). \tag{B.3}$$

This relationship can be inverted to obtain

$$\psi = \psi(R), \tag{B.4}$$

since there is a one-to-one relationship between closed streamlines and the areas they enclose. Differentiating (B.2) with respect to ψ gives

$$\frac{\partial R}{\partial \psi} = \frac{1}{2\pi} \int_{\psi} \frac{\partial x}{\partial \psi} \, dy = \frac{1}{2\pi} \int_{\psi} \frac{\partial \psi}{\partial x}^{-1} \frac{dy}{dt} \, dt = \frac{1}{2\pi} \int_{\psi} dt = \frac{1}{2\pi} T(\psi), \tag{B.5}$$

which gives the time-derivative of the angle θ. Thus, in terms of action-angle coordinates (R, θ) on a region of closed streamlines, the equations for the trajectories take the form

$$\dot{R} = \frac{\partial \psi}{\partial \theta} = 0,$$

$$\dot{\theta} = -\frac{\partial \psi}{\partial R} = -\frac{2\pi}{T(R)} \equiv \Omega(R),$$

where $\Omega(R)$ is the frequency of the motion around the closed streamline with action R.

The reader can refer to the references at the beginning of this appendix for proofs that the Jacobian of the transformation to action-angle coordinates is identically one, which implies that area is preserved under the action-angle transformation, and that the Hamiltonian structure is also preserved under action-angle transformations. This latter point means that if we take the streamfunction in the original (x, y) coordinates and transform it to action-angle coordinates, then the velocity field in action-angle coordinates is obtained from this transformed streamfunction in the usual way. More important, the transformed streamfunction is a function of just the R variable.

For the cellular flow discussed in Section 1.6, action-angle coordinates can be explicitly computed as follows. In the cell $0 \leq x \leq \frac{\pi}{k}$, $0 \leq y \leq 1$, consider a closed trajectory defined by the level set of the streamfunction $\psi_0 = \text{constant} = \frac{A}{k} \sin kx \sin \pi y$.

In order to compute the angle coordinate, it is necessary to compute the time of flight from a point on this trajectory starting from a reference line L to any arbitrary point on the closed trajectory. As is typically the case, this cannot be done globally, since one cannot express x as a function of y, or y as a function of x, along the complete closed trajectory. Consequently, the trajectory must be divided into four arcs as follows (Figure B.3):

Arc 1: $\dot{x} \leq 0$, $\dot{y} \leq 0$, $\frac{\pi}{2k} \leq x \leq x_r$, $y_b \leq y \leq \frac{1}{2}$, where x_r satisfies $\frac{k\psi_0}{A} = \sin kx_r$ and y_b satisfies $\frac{k\psi_0}{A} = \sin \pi y_b$.

Arc 2: $\dot{x} \geq 0$, $\dot{y} \leq 0$, $\frac{\pi}{2k} \leq x \leq x_r$, $\frac{1}{2} \leq y \leq y_t$, where x_r satisfies $\frac{k\psi_0}{A} = \sin kx_r$ and y_t satisfies $\frac{k\psi_0}{A} = \sin \pi y_t$.

Arc 3: $\dot{x} \geq 0$, $\dot{y} \geq 0$, $x_l \leq x \leq \frac{\pi}{2k}$, $\frac{1}{2} \leq y \leq y_t$, where x_l satisfies $\frac{k\psi_0}{A} = \sin kx_l$ and y_t satisfies $\frac{k\psi_0}{A} = \sin \pi y_t$.

Arc 4: $\dot{x} \leq 0$, $\dot{y} \geq 0$, $x_l \leq x \leq \frac{\pi}{2k}$, $y_b \leq y \leq \frac{1}{2}$, where x_l satisfies $\frac{k\psi_0}{A} = \sin kx_l$ and y_b satisfies $\frac{k\psi_0}{A} = \sin \pi y_b$.

The subscripts l, r, t, and b denote left, right, top, and bottom, respectively (Figure B.4).

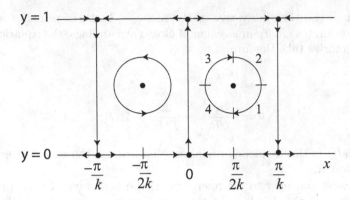

FIGURE B.3. The four arcs of the level set of the streamfunction in the cell $0 \le x \le \frac{\pi}{k}, 0 \le y \le 1$.

Fix a level set of the streamfunction as follows:

$$\sin kx \sin \pi y = \sqrt{1 - \kappa^2} = \frac{k}{A}\psi_0. \tag{B.6}$$

This defines the parameter κ, which satisifies $0 \le \kappa \le 1$. Note that $\kappa = 0$ corresponds to the elliptic stagnation point in the center of the cell and $\kappa = 1$ corresponds to the cell boundary. Moreover, with (B.6), κ can be expressed as a function of ψ_0:

$$\kappa = \sqrt{1 - \left(\frac{k}{A}\psi_0\right)^2}. \tag{B.7}$$

It is necessary also to express z as a function of x, or x as a function of z,

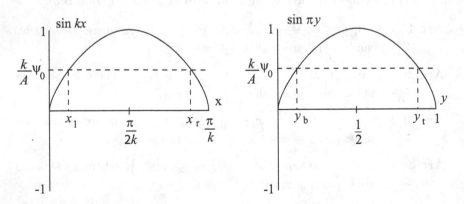

FIGURE B.4. The points x_l, x_r, y_t, and y_b.

on the various arcs of the circle. Towards this end, using (B.6), we have

$$\cos \pi y = \begin{cases} +\sqrt{1 - \frac{\kappa'^2}{\sin^2 kx}}, & 0 \le y \le \frac{1}{2}, \\ \\ -\sqrt{1 - \frac{\kappa'^2}{\sin^2 kx}}, & \frac{1}{2} \le y \le 1, \end{cases} \tag{B.8}$$

$$\cos kx = \begin{cases} +\sqrt{1 - \frac{\kappa'^2}{\sin^2 \pi y}}, & 0 \le x \le \frac{\pi}{2k}, \\ \\ -\sqrt{1 - \frac{\kappa'^2}{\sin^2 \pi y}}, & \frac{\pi}{2k} \le x \le \frac{\pi}{k}, \end{cases} \tag{B.9}$$

where it is convenient to define the new parameter

$$\kappa' = \sqrt{1 - \kappa^2} = \frac{k}{A} \psi_0, \tag{B.10}$$

and it is helpful to note that

$$\sin \pi y_t = \sin \pi y_b = \sin kx_r = \sin kx_l = \kappa'. \tag{B.11}$$

COMPUTATION OF THE ANGLE θ

In this computation, care must be taken to choose the correct branch of the function in each integral and to get the corresponding signs correct. In order to compute the angle transformation (recall (B.1)), it is necessary first to compute the time of flight from a reference point on the closed trajectory to an arbitrary point on that trajectory. This will be done locally on each of the four arcs, and the results will then be combined to get the global description.

From (3.26) and (B.8), it follows that

$$\frac{dx}{dt} = \pm \frac{A\pi}{k} \sqrt{\sin^2 kx - \kappa'^2}. \tag{B.12}$$

Here, take the plus sign on arcs 2 and 3, and the minus sign on arcs 1 and 4.

These equations yield the following expressions of t as a function of x on the four arcs. On arcs 2 and 4,

$$\int_0^t dt' = \pm \frac{k}{A\pi} \int_{\frac{\pi}{2k}}^x \frac{dx'}{\sqrt{\sin^2 kx' - \kappa'^2}}, \tag{B.13}$$

with the plus sign on arc 2 , and the minus sign on arc 4. On arc 1,

$$\int_0^t dt' = -\frac{k}{A\pi} \int_{x_r}^x \frac{dx'}{\sqrt{\sin^2 kx' - \kappa'^2}}, \tag{B.14}$$

and on arc 3,

$$\int_0^t dt' = \frac{k}{A\pi} \int_{x_l}^x \frac{dx'}{\sqrt{\sin^2 kx' - \kappa'^2}}. \qquad (B.15)$$

Now make the change of variables

$$s' = \sin kx',$$

so that

$$ds' = k \cos kx' dx'$$

and

$$dx' = \pm \frac{1}{k} \frac{ds'}{\sqrt{1 - s'^2}},$$

with the plus sign on arcs 3 and 4, and the minus sign on arcs 1 and 2.

After this change of variables, the integrals (B.13), (B.14), and (B.15) become

$$t = \frac{1}{A\pi} \int_{s=\sin kx}^1 \frac{ds'}{\sqrt{1 - s'^2}\sqrt{s'^2 - \kappa'^2}} \qquad (B.16)$$

on arcs 2 and 4, and

$$t = \frac{1}{A\pi} \int_{\kappa'}^{s=\sin kx} \frac{ds'}{\sqrt{1 - s'^2}\sqrt{s'^2 - \kappa'^2}} \qquad (B.17)$$

on arcs 1 and 3.

These integrals can be represented as the inverses of Jacobi elliptic functions. Thus, (B.16) can be written as

$$A\pi t = dn^{-1}[\sin kx, \kappa],$$

or

$$\sin kx = dn\,[A\pi t, \kappa] \qquad \left(\text{note that } x(0) = \frac{\pi}{2k}\right). \qquad (B.18)$$

Similarly, (B.17) can be written as

$$A\pi t = nd^{-1}\left[\frac{\sin kx}{\kappa'}, \kappa\right],$$

or

$$\sin kx = \kappa' nd[A\pi t, \kappa] \qquad (\text{note that } \sin kx(0) = \kappa'). \qquad (B.19)$$

The period can be calculated using either (B.16) or (B.17). Each of these four integrals can be used to find one-fourth of the period by integrating around the full arc. Upon doing this one obtains

$$T(\kappa) = \frac{4}{A\pi} K(\kappa), \qquad (B.20)$$

where $K(\kappa)$ is the complete elliptic integral of the first kind. Note that $K(0) = \frac{\pi}{2}$ and that $K(\kappa) \to \infty$ monotonically as $\kappa \to 1$.

Recall that $\kappa = 0$ corresponds the the stagnation point of center stability type in the middle of the cell. The eigenvalues of the matrix associated with the linearization of the velocity field at this stagnation point are purely imaginary and are given by $\pm i A\pi$. From (B.20), it follows that $T(0) = \frac{2}{A}$, which is the same as 2π divided by the magnitude of the eigenvalue of the matrix associated with the linearization about this stagnation point. With (B.7), the period can be expressed as a function of ψ_0.

Next, compute t as a function of y. From (3.26) and (B.9), it follows that

$$\frac{dy}{dt} = \pm A\sqrt{\sin^2 \pi y - \kappa'^2}, \tag{B.21}$$

with the plus sign on arcs 3 and 4, and the minus sign on arcs 1 and 2.

These equations yield the expressions for t as a function of y on the four arcs:

$$\int_0^t dt' = \pm \frac{1}{A} \int_{\frac{1}{2}}^y \frac{dy'}{\sqrt{\sin^2 \pi y' - \kappa'^2}}, \tag{B.22}$$

with the plus sign on arc 3, and the minus sign on arc 1. On arc 2,

$$\int_0^t dt' = -\frac{1}{A} \int_{y_t}^y \frac{dy'}{\sqrt{\sin^2 \pi y' - \kappa'^2}}, \tag{B.23}$$

and on arc 4,

$$\int_0^t dt' = \frac{1}{A} \int_{y_b}^y \frac{dy'}{\sqrt{\sin^2 \pi y' - \kappa'^2}}. \tag{B.24}$$

Now make the change of variables

$$s' = \sin \pi y',$$

so that

$$dy' = \pm \frac{1}{\pi} \frac{ds'}{\sqrt{1 - s'^2}},$$

with the plus sign on arcs 1 and 4, and the minus sign on arcs 2 and 3.

After this change of variables, the integrals (B.22), (B.23), and (B.24) become

$$t = \frac{1}{A\pi} \int_{\kappa'}^{s=\sin \pi y} \frac{ds'}{\sqrt{1 - s'^2}\sqrt{s'^2 - \kappa'^2}} \tag{B.25}$$

on arcs 2 and 4, and

$$t = \frac{1}{A\pi} \int_{s=\sin \pi y}^1 \frac{ds'}{\sqrt{1 - s'^2}\sqrt{s'^2 - \kappa'^2}} \tag{B.26}$$

on arcs 1 and 3.

These integrals can be represented as the inverse of Jacobi elliptic functions. Thus, (B.25) can be written as

$$A\pi t = nd^{-1}\left[\frac{\sin \pi y}{\kappa'}, \kappa\right],$$

or

$$\sin \pi y = \kappa' nd[A\pi t, \kappa] \qquad (\text{note that } \sin \pi y(0) = \kappa'). \qquad (B.27)$$

Similarly, (B.26) can be written as

$$A\pi t = dn^{-1}[\sin \pi y, \kappa],$$

or

$$\sin \pi y = dn[A\pi t, \kappa] \qquad \left(\text{note that } y(0) = \frac{1}{2}\right). \qquad (B.28)$$

The period can be calculated using either (B.25) or (B.26). Each of these four integrals can be used to find one-fourth of the period by integrating around the full arc. Upon doing this one obtains

$$T(\kappa) = \frac{4}{A\pi} K(\kappa). \qquad (B.29)$$

Finally, recall from (B.1) that the angle variable is defined in terms of the time of flight by

$$\theta(x, y) = \frac{2\pi}{T(\psi_0)} t(x, y), \qquad (B.30)$$

where, on the appropriate arc, $t(x, y)$ is solely a function of x or of y. An explicit expression for the angle is obtained by substituting the appropriate expression for $t(x)$ or $t(y)$ (depending on which arc is chosen) given above into (B.30).

COMPUTATION OF THE ACTION R

Using (B.2), the action of the closed trajectory labeled by ψ_0 is given by

$$R = \frac{1}{2\pi} \oint_{\psi_0} x \, dy. \qquad (B.31)$$

This integral is left as an exercise for the reader.

Appendix C
Numerical Methods

Accurate numerical methods for the computation of hyperbolic trajectories and their stable and unstable manifolds are necessary for the successful application of the Lagrangian transport analysis methods described in this text. The development of such methods has been the subject of much research in the dynamical systems community. Computational techniques for two-dimensional, time-periodic velocity fields have been surveyed by Parker & Chua (1989) and Mancho et al. (2003). Methods applicable to two-dimensional, quasiperiodic velocity fields are described in Beigie et al. (1994) and Malhotra & Wiggins (1998). This latter reference is also relevant to the general two-dimensional aperiodically time-dependent case, but a more refined and widely applicable method is given in Mancho et al. (2003). Methods to compute two-dimensional unstable manifolds of hyperbolic stagnation points in three-dimensional steady flows are described in Krauskopf et al. (2005).

All of these methods have been developed in the context where analytical formulas for the flows are available (i.e., kinematic models). In general, however, it is necessary to consider flows that are defined in terms of discrete numerical data sets, such as flows that are obtained through numerical solution of discretized dynamical equations of motion. The attempt to analyze transport in the dynamical systems framework for such flows raises new mathematical issues. First, this type of dynamical system is not a set of equations, but a data set defined on a space-time grid. Second, in general one would not expect such a general flow to be steady or periodic in time; instead, it will likely be known only for a finite time interval. This raises questions regarding the applicability of dynamical systems ideas, which traditionally describe primarily the asymptotic, long-time behavior of a system. Indeed, the mathematical definitions of hyperbolic trajectories, stable and unstable manifolds of hyperbolic trajectories, KAM tori, and chaos generally refer to the infinite-time limit. If a flow field is known for only a finite time, and is not time-periodic, so that no inference about the behavior outside the known time interval can be made, additional considerations are necessary for the dynamical systems analysis. This is an area of continuing research (e.g., Wiggins (2005)). Here, the central practical issues associated with computing hyperbolic trajectories and their stable and unstable manifolds are briefly discussed.

Hyperbolic Trajectories

In Chapter 4, hyperbolic trajectories were defined using the notion of Lyapunov exponents and exponential dichotomies. Here the definition is given again in terms of Lyapunov exponents, but in a slightly different form that provides a straightforward application to velocity fields known only on a finite time interval. The discussion here will apply to two-dimensional systems.

HYPERBOLIC TRAJECTORIES: INFINITE TIME

As we described in Chapter 4, linearized stability properties of a given trajectory are obtained from properties of the principal fundamental solution matrix, denoted by $X(t,t_1)$, which is normalized so that $X(t_1,t_1) = $ id, where id denotes the 2×2 identity matrix. Consider the following matrix formed from the fundamental solution matrix:

$$M(t,t_1) \equiv X^T(t,t_1)X(t,t_1), \qquad (C.1)$$

where the superscript T denotes the transpose. Clearly this is a symmetric, positive definite matrix, and therefore it has two real eigenvalues. Now consider the following limit:

$$M \equiv \lim_{(t-t_1)\to\infty} \left(X^T(t,t_1)X(t,t_1)\right)^{\frac{1}{2(t-t_1)}}. \qquad (C.2)$$

The *Lyapunov exponents* are the logarithms of the eigenvalues of M (Oseledec (1968)). They have the following interpretation. Consider an infinitesimal circle centered at the starting point of the trajectory. As this circle evolves under the linearized flow about the trajectory it deforms into an ellipsoid. The Lyapunov exponents are the average logarithmic expansion rates (under the flow linearized about the trajectory) of the principal axes of this ellipsoid. One can say that the trajectory is *hyperbolic* if none of its Lyapunov exponents are zero.

HYPERBOLIC TRAJECTORIES: FINITE TIME

The discussion above suggests a natural generalization of the notion of Lyapunov exponents and hyperbolicity to the situation in which the velocity field is defined only over a finite time interval, i.e., $t \in [t_1,t_2]$. As in the infinite-time case, choose a trajectory, linearize about this trajectory, and then solve the linearized equations for the principal fundamental solution matrix. From the principal fundamental solution matrix, form the symmetric, positive definite matrix $M(t,t_1)$ as in (C.1), and now consider instead the finite-time quantity

$$M(t_2,t_1) \equiv \left(X^T(t_2,t_1)X(t_2,t_1)\right)^{\frac{1}{2(t_2-t_1)}}, \qquad (C.3)$$

where $|t_2 - t_1|$ is presumed to be large, in some general sense relating to local convergence rates. The logarithms of the eigenvalues of $M(t_2, t_1)$ can be called *finite-time Lyapunov exponents*. The trajectory can then be called *hyperbolic on the finite time interval* $[t_1, t_2]$ if none of its finite-time Lyapunov exponents (on this time interval) are zero. It is shown in Ide et al. (2002) that if the finite-time Lyapunov exponents are given by $d_1 < 0$ and $d_2 > 0$, then a time-dependent change of coordinates can be computed such that, in the new coordinates, the velocity field linearized about the hyperbolic trajectory has the form

$$\dot{y}_1 = d_1 y_1,$$
$$\dot{y}_2 = d_2 y_2, \qquad \qquad \text{(C.4)}$$

where $d_2 = -d_1$ for incompressible flow. That is, the linearized flow has the form of a standard saddle. This provides motivation for the above definition of finite-time hyperbolicity. The transformed equations (C.4) have the added advantage that the stable and unstable manifolds of the origin are trivially identified. These can be transformed back into the original coordinates to provide the initialization for an algorithm for computing the stable and unstable manifolds of a hyperbolic trajectory developed in Mancho et al. (2003) and Mancho et al. (2004), and which is described below. However, it must be emphasized that this coordinate transformation, and therefore the form of (C.4), is valid only on the finite time interval under consideration where the finite-time Lyapunov exponents are hyperbolic, in the sense described above.

This definition of finite-time Lyapunov exponent has been in use in the predictability community for some time, see Lapeyre (2002), Legras & Vautard (1996), and Farrell & Ioannou (1996). Spatial distributions of finite-time Lyapunov exponents have been used to rapid mixing regions and barriers to transport for some years now. One of the earliest studies is that of Pierrehumbert (1991b) in the context of an atmospheric transport problem. Haller (2000), Haller & Yuan (2000), and Haller (2001) develop methods for computing hyperbolic trajectories and associated material that are related to finite-time Lyapunov exponents and the velocity-gradient tensor.

The attempt to develop practical and rigorous finite-time analogues to infinite-time quantities such as hyperbolicity, and to retain similar terminology, involves subtleties and potential confusions. Classically, hyperbolicity is quantified by a type of infinite-time average over the entire trajectory, as described above. Therefore, in the classical dynamical systems sense, hyperbolicity cannot change along a trajectory. However, the finite-time hyperbolicity defined above is not constrained in the same way. This hyperbolicity criterion could change from time window to time window, as in the examples discussed by Mancho et al. (2006). This kind of variable hyperbolicity might even be anticipated to occur in turbulent flows, where large temporal fluctuations occurred over a broad range of length scales. It

is likely that additional hyperbolicity definitions and concepts will develop
as the range of applications of these methods expands.

Finding Hyperbolic Trajectories

There are a number of numerical methods that can be used to locate saddle-
point behavior in a flow field. The fluid kinematics near a saddle point are
generally robust: patches of fluid near a saddle are rapidly contracted in the
stable direction and expanded in the unstable direction, and so very rapidly
come to resemble segments of the unstable manifold in forward time, and
segments of the stable manifold in reversed time. Taking the intersection
of these two segments at the appropriate (same) time will identify a small
region containing the hyperbolic trajectory.

A region in which a patch of fluid contracts in one direction and stretches
in another as it evolves in time with the flow can be called a *hyperbolic region*. The rapid stretching and contraction in these hyperbolic regions of
the flow form the foundation of a variety of numerical techniques for finding
hyperbolic trajectories and their stable and unstable manifolds. However,
such hyperbolic regions must themselves be identified before these techniques can be used. Several approaches have been taken to identify these
regions. In one such approach, the trajectories are analyzed for a steady
velocity field that is obtained from the time-dependent field by fixing time
at a particular instant. This is called a "frozen-field" approximation, and a
stagnation point of the frozen field is called a "frozen-time" or "instantaneous" stagnation point (ISP). If the ISP is a saddle point, then it may be
a useful guess for a hyperbolic region in the original, time-dependent flow.
If such a region can be found, more refined techniques can then be applied
to identify hyperbolic trajectories. Examples of the application of this approach can be found in Beigie et al. (1994), Malhotra & Wiggins (1998),
Miller et al. (1997), Rogerson et al. (1999), Yuan et al. (2001), Yuan et al.
(2004), and Poje et al. (2002).

Of course, ISPs are not, in general, stagnation points or even trajectories
of the original, time-dependent velocity field, and there are situations in
which trying to infer trajectory information from this approach can lead to
contradictory results. Ide et al. (2002) present examples of this, as well as
a discussion of the relation between curves of ISPs and fluid trajectories.
However, in situations in which the time variation of the velocity field is
sufficiently slow, curves of ISPs may stay close to a true trajectory of the
vector field. This is intuitively clear from perturbation theory, and therefore, it is not surprising that such results have been used in many fields,
often with no knowledge of the previous history. For example, applications of such results can be found in control theory; see Desoer (1969) and
also Peuteman et al. (2000), which contains historical references for control
theory applications. In the fluid-mechanical context of Stokes flows, such

results were used in Kaper (1992). Recently, these results were further extended in Haller & Poje (1998), with an application to cross-stream mixing in a double-gyre ocean model given in Poje & Haller (1999). In the purely mathematical context, related results can be found in Coppel (1978). ISPs are a prominent features in Eulerian velocity fields, and can be useful indicators of Lagrangian motion. However, the relationship between ISPs and Lagrangian motion is, in general, extremely complex, as examples in Ide et al. (2002) and Mancho et al. (2006) illustrate.

Ide et al. (2002), Ju et al. (2003), and Mancho et al. (2004) introduce a method for locating hyperbolic trajectories that does not require slow time-variation or a priori knowledge of the location of hyperbolic regions. This method is based on an iteration technique in a space of curves parametrized over the (temporal) length of the velocity field, which, when it converges, is guaranteed to converge to a fluid trajectory. Hyperbolicity is checked during each step of the convergence process. The method uses hyperbolic ISPs as an initial guess in the iterative process (although other "good" guesses are also possible), but this is not equivalent to an assumption of hyperbolicity for the trajectories of the velocity field since ISPs are not trajectories. Moreover, the curves of ISPs need not remain close to the hyperbolic trajectories that are located by a successful iteration. The main innovations of this method are that it provides an approximation to a hyperbolic trajectory for the entire time interval on which the velocity field is defined, and that this approximation is itself a trajectory (rather than a thin region or group of points). These are significant advantages over other, simpler methods.

Computation of Material Manifolds

After a hyperbolic trajectory has been located, the next step in the analysis is to compute its stable and unstable manifolds. The difficulty of this step should not be underestimated, since many types of flows encountered in geophysical fluid dynamics are different, and more complex, than those typically considered in dynamical systems theory. For example, the nature of the time-dependence may vary significantly depending on the spatial coordinates. Also, the aperiodicity in time gives rise to more intensive computational requirements for processor time, memory, and storage. These issues are addressed and discussed in the series of papers described below. In Mancho et al. (2004) the hyperbolic trajectory algorithm of Ide et al. (2002) and Ju et al. (2003) is further developed and merged with the stable and unstable manifold computation algorithm described in Mancho et al. (2003) into a single, unified algorithm.

These manifold-approximation algorithms generally involve two stages. First, the linear stable and unstable subspaces $E^s(t)$ and $E^s(t)$ are computed for the hyperbolic trajectory. Then, the stable and unstable material manifolds $W^s(t)$ and $W^u(t)$ are obtained by computing trajectories with initial conditions close to the hyperbolic trajectory and on the linear subspaces. These trajectories are computed in forward time for the unstable manifold, and in backward time for the stable manifold, from a small initial segment of the manifold with one endpoint at the hyperbolic trajectory. In each case, the forward and backward integration intervals are chosen so that the computed manifolds are obtained at the same time.

In the method described in Mancho et al. (2004), the initial segment is obtained from the algorithm that is used to locate the hyperbolic trajectory, following a numerical transformation to the standard saddle (C.4). The short initial segments consist of closely spaced points along the linear subspaces. The subsequent forward and backward integrations that extend these initial segments along the manifolds can cause gaps to develop as the points separate, and, sometimes, accumulation of points in other regions. A detailed study of a variety of numerical approaches to maintaining adequate and approximately uniform resolution along the manifolds is carried out in Mancho et al. (2003), where it is demonstrated that sophisticated techniques developed by Dritschel (1989) and Dritschel & Ambaum (1997) in another context can be used for interpolation, controlling the size of gaps between points, and point redistribution along the computed manifolds. These numerical techniques were developed for the study of the evolution of the boundaries of vortex patches in complex fluid flows, and are ideally suited for computing the evolution of material curves in general flows. This approach to computing hyperbolic trajectories and their stable and unstable manifolds has been used for a complex, time-dependent, wind-driven double-gyre flow by Mancho et al. (2004).

Appendix D

Finite-Time Material Manifolds: An Example

This example is taken from Mancho et al. (2006). It illustrates several properties of the stable and unstable manifolds of unsteady, nonlinear velocity fields that are defined only for a finite time.

Consider the velocity field

$$\dot{x} = x,$$

$$\dot{y} = -y + x^2 \left(\frac{1}{3}\dot{a}(t) + a(t) \right), \tag{D.1}$$

where $a(t)$ is an arbitrary, time-dependent function that is defined over a finite time interval. For example,

$$a(t) = (t(t-1))^{\frac{3}{2}} \quad \text{for} \quad 0 \le t \le 1. \tag{D.2}$$

The velocity field (D.1) can be derived from the streamfunction

$$\psi(x, y) = -xy + \frac{x^3}{3} \left(\frac{1}{3}\dot{a}(t) + a(t) \right). \tag{D.3}$$

It is simple to verify that $(x, y) = (0, 0)$ is a trajectory of (D.1). In general, of course, it is unusual to find stagnation point trajectories in unsteady flows, but this simplified situation is convenient for the present purposes.

Linearization about the trajectory $(x, y) = (0, 0)$ yields the velocity field

$$\begin{pmatrix} \dot{\xi} \\ \dot{\eta} \end{pmatrix} = \begin{pmatrix} 1 & 0 \\ 0 & -1 \end{pmatrix} \begin{pmatrix} \xi \\ \eta \end{pmatrix}. \tag{D.4}$$

The stable and unstable subspaces of the linearized velocity field are given by the η axis and ξ axis, respectively. That these subspaces are time-independent is a second convenient simplification that generally would not hold in unsteady flows, even for stagnation-point trajectories.

The form of the linearized equations (D.4) is consistent with the identification of the stagnation point $(x, y) = (0, 0)$ as a hyperbolic trajectory, as defined in terms of exponential dichotomies in Section 4.6. A subtle point is at issue here: generally, one would anticipate that the definition of hyperbolicity in terms of exponential dichotomies would require that the asymptotic behavior of the flow as $t \to \pm\infty$ be known, since the exponential bounds (4.30) and (4.31) are technically trivial on finite time intervals (that is, suitable constants can always be found so that the inequalities are satisfied on finite intervals). On the other hand, the original definition of exponential dichotomies by Coppel (1978) refers to a finite time interval. Some theoretical approaches to the definition of hyperbolicity in finite-time flows are discussed in Section 7.5. From a practical point of view, the identification of the finite-time trajectory as hyperbolic will make sense if the time scale of the exponential behavior of the linearized flow (D.4) is sufficiently short relative to the time interval on which the flow is defined. It is assumed that this is the case in the present example, and that the stagnation point can therefore be sensibly identified as a hyperbolic trajectory.

Next, consider the nonlinear velocity field. It is easy to verify that on $x = 0$, we have $\dot{x} = 0$ in (D.1). Hence the y axis is the *global* stable manifold of the hyperbolic trajectory $(x, y) = (0, 0)$. The global unstable manifold is given by

$$y = \frac{a(t)}{3} x^2. \tag{D.5}$$

This curve passes through the origin, and is tangent to the unstable subspace of the linearized velocity field at the origin, for all time that the velocity field (i.e., $a(t)$) is defined. Invariance of the curve (D.5) means that the velocity is tangent to the curve. In order to check this invariance and verify that (D.5) indeed gives the unstable manifold, it is necessary to differentiate (D.5) with respect to time,

$$\dot{y} = \frac{\dot{a}(t)}{3} x^2 + \frac{2a(t)}{3} x\dot{x}, \tag{D.6}$$

and show that $\dot{x}$ and $\dot{y}$ from (D.1) satisfy (D.6). This is a simple calculation that is left to the reader.

There are two main points to be taken from this example:

- Even if the velocity field exists for only a finite time interval, it may be possible to define a unique, distinguished hyperbolic trajectory whose stable and unstable manifolds divide the flow into qualitatively distinct regions, which describe qualitatively different evolutions of the trajectories.

- Even if the velocity field exists for only a finite time interval, at each instant of time the stable and unstable manifolds may be infinite in length, as is the curve (D.5).

Note that illustrations of manifolds in diagrams necessarily show curves of finite length. However, each stable and unstable manifold is generally of infinite length. The small segment shown in an illustration is analogous to a streak of dye that is injected over a finite time interval and is stretched along the manifold. In such laboratory experiments, the hyperbolic (i.e., exponential) structure rapidly contracts the dye toward the manifold and stretches it out along the manifold. This process and point of view are described in some detail in Beigie et al. (1994), and their description explains why only unstable manifolds are seen in laboratory flow visualizations with dye or other tracers: it would be necessary to make time run backwards in order to see the stable manifold.

Glossary

Area-Preserving

An area-preserving flow is a two-dimensional flow that is also *incompressible*.

Autonomous

See *steady flow*.

Baroclinic Flows

A baroclinic flow is one in which pressure is not constant on surfaces of constant density. See, for example, Pedlosky (1987), Salmon (1998), or Holton (1992). In general, baroclinic flows are intrinsically three-dimensional.

Barotropic Flows

A barotropic flow is one in which pressure is constant on surfaces of constant density. See, for example, Pedlosky (1987), Salmon (1998), or Holton (1992). In general, an inviscid barotropic flow is effectively two-dimensional.

Critical Lines, Levels, Layers

A critical line is a line $y = y_c$ in a horizontally varying parallel shear flow, at which the phase speed c of a disturbance is equal to the local flow velocity $U(y_c)$. A critical level is a line $z = z_c$ in a vertically varying parallel shear flow, at which the phase speed c of a disturbance is equal to the local flow velocity $U(z_c)$. A critical layer may be either a critical line or level, or the region immediately surrounding and including a critical line or level in which the flow becomes nonlinear even for small-amplitude disturbances. Critical lines play an important dynamical role in the linear theory of fluid-dynamical instabilities (e.g., Drazin & Reid (1981), Pedlosky (1987)).

Divergence-Free

A divergence-free velocity field is *incompressible*.

Equilibrium Point

See *stagnation point*.

Fixed Point

See *stagnation point.*

Geostrophic

A geostrophic flow is a rotating flow in which the first-order balance in the horizontal momentum equations is between the Coriolis and pressure gradient forces. See, for example, Pedlosky (1987), Salmon (1998), or Holton (1992). On horizontal scales greater than the ocean depth, and time scales larger than a day, most ocean motions are approximately geostrophic.

Gulf Stream

The Gulf Stream is a persistent western boundary current or *jet* in the North Atlantic Ocean, with a nominal width of 100 km and a volume flux of 30 to 100×10^6 m^3 s^{-1}. It flows northward along the east coast of North America, and then eastward into the interior of the North Atlantic north of Cape Hatteras. The book by Stommel (1965) contains a readable and still useful introduction to observations and simple models of the Gulf Stream.

Heteroclinic Trajectory or Orbit

Let $x_0(t)$ and $x_1(t)$ be two trajectories in a flow. Then a trajectory that approaches $x_0(t)$ as $t \to -\infty$ and $x_1(t)$ as $t \to +\infty$ is called a *heteroclinic trajectory*. Note that the heteroclinic trajectory is in the intersection of $W^u[(x_0(t)]$ and $W^s[(x_1(t)]$, where $W^u[x_0(t)]$ and $W^s[x_1(t)]$ are the unstable and stable manifolds of $x_0(t)$ and $x_1(t)$, respectively.

Homoclinic Trajectory or Orbit

Let $x_0(t)$ be a trajectory in a flow. Then a trajectory that approaches $x_0(t)$ both as $t \to -\infty$ and as $t \to +\infty$ is called a *homoclinic trajectory*. Note that the homoclinic trajectory is in the intersection of $W^u[(x_0(t)]$ and $W^s[(x_0(t)]$, where $W^u[x_0(t)]$ and $W^s[x_0(t)]$ are the unstable and stable manifolds of $x_0(t)$, respectively.

Incompressible Flow

An incompressible flow is a flow with zero divergence, that is, a velocity field **v** satisfying (1.8). From a physical point of view, incompressibility of a flow follows from the principle of mass conservation and the requirement that fluid density either be uniform or remain constant along trajectories. In the geophysical context, the two-dimensional form of (1.8) is also satisfied by any horizontal *geostrophic* velocity field on scales small relative to the Earth's radius. As shown in Appendix A, (1.8) implies that the area of a specific region of fluid in a two-dimensional flow, or the volume in a three-dimensional flow, is unchanged during its evolution under the flow, when the flow is time-dependent.

Jet

A local region of strong, coherent, approximately unidirectional flow within a larger fluid domain. Generally, jet regions are anisotropic, with larger dimensions along than across the flow.

Morse Lemma

In Section 2.3, we described the trajectories near stagnation points for the linearized velocity field, and concluded that a special set of material lines existed, the invariant subspaces E^u and E^s, which divided the trajectories of the linearized velocity field into different flow regimes. Are the trajectories of the linearized velocity field qualitatively accurate descriptions of the trajectories of the full velocity field? For two-dimensional, steady velocity fields, the streamfunction representation and a classical theorem from differential geometry enable us to answer this question.

Recall that in this case the velocity field can be written in the form

$$\dot{x} = -\frac{\partial \psi}{\partial y}(x, y),$$

$$\dot{y} = \frac{\partial \psi}{\partial x}(x, y). \tag{D.7}$$

Now, as above, let $\mathbf{x}_0$ denote a stagnation point. A simple calculation shows that the eigenvalues of the matrix associated with the linearization of the velocity field about $\mathbf{x}_0$ are given by

$$\pm \sqrt{\left(\left(\frac{\partial^2 \psi}{\partial x \partial y} \right)^2 - \frac{\partial^2 \psi}{\partial x^2} \frac{\partial^2 \psi}{\partial y^2} \right)} \bigg|_{\mathbf{x}_0}. \tag{D.8}$$

Therefore, for

$$\left(\frac{\partial^2 \psi}{\partial x \partial y} \right)^2 \bigg|_{\mathbf{x}_0} < \left(\frac{\partial^2 \psi}{\partial x^2} \frac{\partial^2 \psi}{\partial y^2} \right) \bigg|_{\mathbf{x}_0}, \tag{D.9}$$

the stagnation point is a center, and if

$$\left(\frac{\partial^2 \psi}{\partial x \partial y} \right)^2 \bigg|_{\mathbf{x}_0} > \left(\frac{\partial^2 \psi}{\partial x^2} \frac{\partial^2 \psi}{\partial y^2} \right) \bigg|_{\mathbf{x}_0}, \tag{D.10}$$

the stagnation point is a saddle.

Now expand the streamfunction in a Taylor series about $\mathbf{x}_0$:

$$\psi(x_0 + \xi, y_0 + \eta) = \psi(\mathbf{x}_0) + \frac{1}{2} \frac{\partial^2 \psi}{\partial x^2}(\mathbf{x}_0)\xi^2 + \frac{\partial^2 \psi}{\partial x \partial y}(\mathbf{x}_0)\xi\eta$$

$$+ \frac{1}{2} \frac{\partial^2 \psi}{\partial y^2}(\mathbf{x}_0)\eta^2 + \mathcal{O}(3), \tag{D.11}$$

where $\mathcal{O}(3)$ represents polynomial terms in ξ and η of third and higher order. If the determinant of the Hessian matrix of $\psi(\mathbf{x})$ evaluated at $\mathbf{x} = \mathbf{x}_0$ is nonzero, then the Morse lemma (Milnor [1963]) tells us that in a neighborhood of $\mathbf{x}_0$ there exists a nonlinear, differentiable, invertible change of coordinates,

$$\bar{\mathbf{x}} = [\bar{x}(x, y), \bar{y}(x, y)],$$

in which all of the terms of third and higher order in the expansion (D.11) are zero, and the quadratic terms take a particularly simple form. The determinant of the Hessian is nonzero provided that

$$\left. \left(\left(\frac{\partial^2 \psi}{\partial x \partial y} \right)^2 - \frac{\partial^2 \psi}{\partial x^2} \frac{\partial^2 \psi}{\partial y^2} \right) \right|_{\mathbf{x}_0} \neq 0,$$

for which there are two possibilities: (D.9) or (D.10). For the case in which $\mathbf{x}_0$ is a center, the Morse lemma implies that near $\mathbf{x}_0$ there is a coordinate system in which the streamfunction takes the form

$$\psi(\mathbf{x}) = \bar{\psi}(\bar{\mathbf{x}}) = \psi(\mathbf{x}_0) + \bar{x}^2 + \bar{y}^2, \tag{D.12}$$

or

$$\psi(\mathbf{x}) = \bar{\psi}(\bar{\mathbf{x}}) = \psi(\mathbf{x}_0) - \bar{x}^2 - \bar{y}^2. \tag{D.13}$$

So if the stagnation point is a center by the linear criterion, it is also surrounded by closed trajectories of the full velocity field (Figure 2.3b).

For the case in which $\mathbf{x}_0$ is a saddle, the Morse lemma implies that near $\mathbf{x}_0$ there is a coordinate system in which the streamfunction takes the form

$$\psi(\mathbf{x}) = \bar{\psi}(\bar{\mathbf{x}}) = \psi(\mathbf{x}_0) + \bar{x}^2 - \bar{y}^2, \tag{D.14}$$

or

$$\psi(\mathbf{x}) = \bar{\psi}(\bar{\mathbf{x}}) = \psi(\mathbf{x}_0) - \bar{x}^2 + \bar{y}^2. \tag{D.15}$$

So, if the stagnation point is a saddle point by the linearized criterion, it is also a saddle point with regard to the trajectories of the full velocity field (Figure 2.3a).

Nonautonomous

See *unsteady flow*.

Nonresonance Condition (KAM Theorem)

The nonresonance condition (3.36) is referred to as a *Diophantine condition* and the frequency vector $(\Omega(r_0), \omega_1, \ldots, \omega_n)$ is referred to as a *Diophantine frequency vector*. This condition is subtle, since it is actually an infinite number of inequalities that must be satisfied by $(\Omega(r_0), \omega_1, \ldots, \omega_n)$. Such a condition would seem to be "uncheckable," given a specific $(\Omega(r_0), \omega_1, \ldots, \omega_n)$. In some sense it is. Its main value lies in the fact that it is an ideal condition for use in proving the convergence of certain series that arise in proofs of the KAM theorem. Nevertheless, one may ask how it can be used in practice. The key feature here is that "most" frequency vectors are Diophantine. Let us make this notion more precise. Let $[r_{min}, r_{max}]$ denote the action values for the range of closed streamlines of the unperturbed flow that we are considering. Then with the nondegeneracy condition (3.35) satisfied, there is a corresponding interval of frequency vectors $[\Omega(r_{min}), \Omega(r_{max})] \equiv \mathcal{I}$. Let $\mathcal{D}$ denote an open subset in $\mathbb{R}^n$ that represents the domain of all possible frequencies $\omega_1, \ldots, \omega_n$. Let $\mu(\mathcal{I} \times \mathcal{D})$ denote the Lebesgue measure of the set $\mathcal{I} \times \mathcal{D}$. Then the Lebesgue measure of the set of Diophantine vectors in $\mathcal{I} \times \mathcal{D}$ is equal to $\mu(\mathcal{I} \times \mathcal{D})$. In other words, if we were to choose at random a frequency vector from $\mathcal{I} \times \mathcal{D}$ with probability one it would be Diophantine. Proofs of these statements can be found in Wiggins (2003).

Potential Vorticity

The potential vorticity is a scalar *vorticity* invariant, or approximate invariant, that is constructed from the dot product of the vorticity and the gradient of a materially conserved, or approximately conserved, scalar field. Certain potential vorticities, including the *quasigeostrophic* potential vorticity, play especially important roles in geophysical fluid flows. See, for example, Pedlosky (1987), Salmon (1998), or Holton (1992).

Quasigeostrophic

A quasigeostrophic flow is a flow described by one of several varieties of quasigeostrophic potential vorticity equation, which can be derived as approximations for rotating flows that are close to a state of *geostrophic* balance. See, for example, Pedlosky (1987), Salmon (1998), or Holton (1992). Quasigeostrophic models provide useful representations of many large-scale oceanic and atmospheric flows.

Quasiperiodic Time-Dependence

The perturbation $\psi_1(x, y, t, \varepsilon)$ is said to depend quasiperiodically on time t with basic frequencies ω_i, $i = 1, \ldots, n$, if $\psi_1(x, y, t, \varepsilon) = \Psi_1(x, y, \theta_1, \ldots, \theta_n, \varepsilon)$, where Ψ_1 is 2π-periodic in each θ_i, $i = 1, \ldots, n$, and $\theta_i = \omega_i t$, $i = 1, \ldots, n$, where $n \geq 1$ and the ω_i are n incommensurate, constant frequencies.

Rest Point

See *stagnation point*.

Simply Connected

See *Stokes's Theorem*.

Singular Point

See *stagnation point*.

Stable and Unstable Manifold Theorem for Hyperbolic Trajectories

We have described earlier that if we linearize a velocity field about a hyperbolic trajectory, then the associated linearized velocity field has time-dependent stable and unstable subspaces, $E^s(t)$ and $E^u(t)$. These subspaces correspond to initial conditions of trajectories that decay toward the hyperbolic trajectory at an exponential rate as $t \to \infty$ and $t \to -\infty$, respectively. Moreover, these stable and unstable subspaces are time-dependent material lines in the linearized flow and hence act as flow barriers.

Essentially the same situation holds for the nonlinear velocity field. In particular, there exist time-dependent material curves passing through the hyperbolic trajectory that are called the *stable and unstable manifolds of the hyperbolic trajectory*. The stable (respectively, unstable) manifold is tangent to the stable (respectively, unstable) subspace at the hyperbolic trajectory. At a fixed time, the stable (respectively, unstable) manifold corresponds to initial conditions of trajectories that approach the hyperbolic trajectory at an exponential rate as $t \to \infty$ (respectively, $t \to -\infty$). Moreover, they are time-dependent material curves and therefore act as barriers to transport. It is assumed here that $\mathbf{v}(\mathbf{x}, t)$ is continuous in t, for all $t \in \mathbb{R}$, and C^r in $\mathbf{x}$, $r \geq 1$, and, moreover, that all of the partial derivatives of $\mathbf{v}(\mathbf{x}, t)$, with respect to $\mathbf{x}$, up to order r, are uniformly continuous and uniformly bounded in $K \times \mathbb{R}$, where K is an arbitrary closed and bounded subset of the spatial domain $U \subset \mathbb{R}^n$, $n = \{2, 3\}$. These regularity assumptions on the velocity field will be sufficient for the invariant-manifold theorems described below; see Yi (1993) or Kaper (1992).

This result can be stated as a theorem, given here for the two-dimensional case ($n = 2$). At time τ, let $D_\rho[\mathbf{x}_0(\tau)]$ denote the ball of radius ρ centered at the hyperbolic trajectory $\mathbf{x}_0(\tau)$. Then the *local stable and unstable manifold theorem for hyperbolic trajectories* states that for the set-up and hypotheses described above, there exists in $D_\rho[\mathbf{x}_0(\tau)]$ a one-dimensional manifold, $W^s_{\mathrm{loc}}[\mathbf{x}_0(\tau)]$, and a one-dimensional manifold $W^u_{\mathrm{loc}}[\mathbf{x}_0(\tau)]$, such that for $0 < \rho < \rho_0$, with ρ_0 sufficiently small:

1. $W^s_{\mathrm{loc}}[\mathbf{x}_0(\tau)]$ and $W^u_{\mathrm{loc}}[\mathbf{x}_0(\tau)]$ are material curves; in mathematical terms, this means they are invariant under time evolution of the flow, in the sense that trajectories starting on these curves will remain on them.

2. Trajectories starting on $W^s_{\mathrm{loc}}[\mathbf{x}_0(t)]$ at time $t = \tau$ approach $\mathbf{x}_0(t)$ at an exponential rate $e^{-\lambda'(t-\tau)}$ as $t \to \infty$ and trajectories starting on $W^u_{\mathrm{loc}}[\mathbf{x}_0(t)]$ at time $t = \tau$ approach $\mathbf{x}_0(t)$ at an exponential rate $e^{-\lambda'|t-\tau|}$ as $t \to \infty$, for some constant $\lambda' > 0$.

$W_{\text{loc}}^s[\mathbf{x}_0(t)]$ is called the local stable manifold of $\mathbf{x}_0(t)$, and $W_{\text{loc}}^u[\mathbf{x}_0(t)]$ the local unstable manifold of $\mathbf{x}_0(t)$.

In some sense this theorem has been known for some time, although the autonomous version is much more widely known. The theorem can be obtained from simple modifications of results found in Coddington & Levinson (1955) and Hale (1980). The theorem in this form can be found in the Ph.D. thesis of Kaper (1992), see also Yi (1993). A discrete-time version can be found in Katok & Hasselblatt (1995), but see also de Blasi & Schinas (1973) and Irwin (1973).

Some additional properties of the stable and unstable manifolds of a hyperbolic trajectory are as follows:

- The stable and unstable manifolds are C^r curves. This is particularly useful for approximating the manifolds by Taylor expansions.

- $W_{\text{loc}}^s[\mathbf{x}_0(t)]$ and $W_{\text{loc}}^u[\mathbf{x}_0(t)]$ intersect along $\mathbf{x}_0(t)$, and the angle between the manifolds is bounded away from zero uniformly for all $t \in \mathbb{R}$.

- Every trajectory on $W_{\text{loc}}^s[\mathbf{x}_0(t)]$ can be continued to the boundary of $D_\rho[\mathbf{x}_0(t')]$ in backward time for some $t' < t$, and every trajectory on $W_{\text{loc}}^u[\mathbf{x}_0(t)]$ can be continued to the boundary of $D_\rho[\mathbf{x}_0(t'')]$ in forward time for some $t'' > t$.

- Any trajectory in $D_\rho[\mathbf{x}_0(t)]$ *not* on either $W_{\text{loc}}^s[\mathbf{x}_0(t)]$ or $W_{\text{loc}}^u[\mathbf{x}_0(t)]$ will leave $D_\rho[\mathbf{x}_0(t^*)]$ for some positive and negative t^*.

The global (in space and time) stable and unstable manifolds $W^s[\mathbf{x}_0(t)]$ and $W^u[\mathbf{x}_0(t)]$ can be obtained by following trajectories in $W_{\text{loc}}^s[\mathbf{x}_0(t)]$ and $W_{\text{loc}}^u[\mathbf{x}_0(t)]$ backward and forward in time, respectively (Figure 4.2). This is essentially the basis of numerical methods used to obtain numerical representations of the stable and unstable manifolds (Appendix C).

In general, the stable and unstable manifolds $W^s(t)$ and $W^u(t)$ of a hyperbolic trajectory $\mathbf{x}_0(t)$, and the associated linear subspaces $E^s(t)$ and $E^u(t)$, all depend both on time t and on the trajectory $\mathbf{x}_0(t)$. It can be useful to indicate just one of these dependencies explicitly in the notation, and suppress the other, or to indicate both explicitly. Thus, the different notations $W^s(t)$, $W^s[\mathbf{x}_0(t)]$, $W^s[\mathbf{x}_0]$, and $W^s[t, \mathbf{x}_0(t)]$ can all denote the same time-dependent and trajectory-dependent material manifold, but emphasize different aspects of these dependencies, as appropriate in each context. The same is true of the corresponding notations for W^u, E^s, and E_u.

Stable and Unstable Manifold Theorem for Hyperbolic Trajectories: Persistence Under Perturbation

One of the nice features of hyperbolic trajectories and their stable and unstable manifolds is that they persist under perturbation. That is, they are slightly altered, but not destroyed, if the velocity field is changed slightly.

Consider the velocity field

$$\dot{\mathbf{x}} = \mathbf{v}(\mathbf{x}, t; \varepsilon), \tag{D.16}$$

where ε is a parameter or vector of parameters, and

$$\mathbf{v}(\mathbf{x}, t; 0) = \mathbf{v}(\mathbf{x}, t).$$

As before, assume that the velocity field depends sufficiently smoothly on the parameter(s).

Suppose that at $\varepsilon = 0$, there exists a hyperbolic trajectory with local stable and unstable manifolds for (D.16) as in the previous theorem. Then the *persistence theorem for hyperbolic trajectories and stable and unstable manifolds* states that there exists $\varepsilon_0 > 0$ sufficiently small such that for $|\varepsilon| \leq \varepsilon_0$, (D.16) possesses a hyperbolic trajectory $\mathbf{x}_\varepsilon(t)$, having a one-dimensional stable manifold $W^s[\mathbf{x}_\varepsilon(t)]$ and a one-dimensional unstable manifold $W^u[\mathbf{x}_\varepsilon(t)]$. These manifolds have the same properties as those described in the previous theorem, with $\mathbf{x}_0(t)$ replaced by $\mathbf{x}_\varepsilon(t)$. Moreover, they depend smoothly on ε, and the rates of convergence of the corresponding trajectories to $\mathbf{x}_\varepsilon(t)$ are continuous functions of ε. For a proof of this theorem see Kaper (1992) or Yi (1993).

Stagnation Point

A stagnation point is a point where the velocity vanishes. Synonyms for stagnation point in dynamical systems theory include *equilibrium point, fixed point, stationary point, rest point*, and *singular point*.

Stationary Point

See *stagnation point*.

Steady Flow

A simple but important type of flow is *steady* flow, in which the velocity field $\mathbf{v}$ is independent of time, i.e. $\mathbf{v} = \mathbf{v}(\mathbf{x})$. In this case, the equations for the trajectories are said to be *autonomous*.

Stokes's Theorem for Simply and Multiply Connected Regions

Stokes's theorem in two dimensions is also known as Green's theorem in the plane.

Let C denote the boundary of a region $\mathcal{R}$ and let $P(x, y)$ and $Q(x, y)$ denote functions that are continuous and have continuous partial derivatives in $\mathcal{R}$ and on its boundary. Then Green's theorem in the plane states that

$$\oint_C P\, dx + Q\, dy = \iint_{\mathcal{R}} \left(\frac{\partial Q}{\partial x} - \frac{\partial P}{\partial y} \right) dx\, dy.$$

This theorem is valid for both simply-connected and multiply-connected regions. A region, $\mathcal{R}$, is said to be *simply-connected* if any simple closed curve in $\mathcal{R}$ can be shrunk continuously to a point without leaving $\mathcal{R}$. A closed curve that does not intersect itself anywhere is called a *simple closed curve*. If $\mathcal{R}$ is not simply-connected then it is called *multiply-connected*.

Sverdrup

A Sverdrup is a unit of volume flux equal to 10^6 m^3s^{-1}, which is convenient for measuring the volume transport of ocean currents.

Unsteady Flow

A flow is said to be unsteady if it is not *steady*, that is, if the velocity field depends explicitly on time: $\mathbf{v} = \mathbf{v}(\mathbf{x}, t)$. In this case, the equations for the trajectories are said to be *nonautonomous*, that is, not *autonomous*.

Vorticity

The vorticity is the curl of the velocity field, and is a measure of the local angular momentum of the fluid. See, for example, Pedlosky (1987), Salmon (1998), or Holton (1992).

Western Boundary Current

Western boundary currents, such as the Gulf Stream in the North Atlantic, are strong currents along the western boundaries of ocean basins that are prominent features of all mid-latitude ocean gyres. See, for example, Pedlosky (1987) or Salmon (1998).

References

Acrivos, A., Aref, H., & Ottino, J. M., editors (1991). *Symposium on Fluid Mechanics of Stirring and Mixing*, volume 3(5) of *Phys. Fluids A, Part 2*.

Aref, H. (2002). The development of chaotic advection. *Phys. Fluids*, **14(4)**, 1315–1325.

Arnold, V. I. (1963). Proof of A. N. Kolmogorov's theorem on the preservation of quasiperodic motions under small perturbations of the Hamiltonian. *Russ. Math. Surveys*, **18(5)**, 9–36.

Arnold, V. I. (1973). *Ordinary Differential Equations*. M.I.T. Press, Cambridge, MA.

Arnold, V. I. (1989). *Mathematical Methods of Classical Mechanics*. Springer-Verlag, New York, Heidelberg, Berlin.

Arnold, V. I., Kozlov, & Neishtadt, A. I. (1988). *Mathematical Aspects of Classical and Celestial Mechanics*, volume III of *Dynamical Systems*. Springer-Verlag, New York, Heidelberg, Berlin.

Aubry, S. (1983a). Devil's staircase and order without periodicity in classical condensed matter. *J. Physique*, **44**, 147–162.

Aubry, S. (1983b). The twist map, the extended Frenkel–Kontorova model and the devil's staircase. *Physica D*, **7**, 240–258.

Babiano, A., Provenzale, A., & Vulpiani, A., editors (1994). *Chaotic Advection, Tracer Dynamics, and Turbulent Dispersion. Proceedings of the NATO Advanced Research Workshop and EGS Topical Workshop on Chaotic Advection, Conference Centre Sereno di Gavo, Italy, 24–28 May 1993*, volume 76 of *Physica D*.

Balasuriya, S. (2001). Gradient evolution for potential vorticity flows. *Nonlinear Processes in Geophysics*, **8(4-5)**, 253–263.

Bartlett, J. H. (1982). Limits of stability for an area-preserving polynomial mapping. *Celest. Mech.*, **28**, 295–317.

Batchelor, G. K. (1967). *An Introduction to Fluid Dynamics*. Cambridge University Press, Cambridge.

Beigie, D., Leonard, A., & Wiggins, S. (1991). Chaotic transport in the homoclinic and heteroclinic tangle regions of quasiperiodically forced two-dimensional dynamical systems. *Nonlinearity*, **4**, 775–819.

Beigie, D., Leonard, A., & Wiggins, S. (1994). Invariant manifold templates for chaotic advection. *Chaos, Solitons, and Fractals*, **4(6)**, 749–868.

Bennett, A. F. (2006). *Lagrangian Fluid Dynamics*. Cambridge Monographs on Mechanics. Cambridge University Press, Cambridge.

Boffetta, G., Lacorata, G., Redaelli, G., & Vulpiani, A. (2001). Detecting barriers to transport: a review of different techniques. *Physica D*, **159**, 58–70.

Bolton, E. W., F. Busse, F., & Clever, R. M. (1986). Oscillatory instabilities of convection rolls at intermediate Prandtl numbers. *J. Fluid Mech.*, **164**, 469–486.

Bower, A. S. (1991). A simple kinematic mechanism for mixing fluid parcels across a meandering jet. *J. Phys. Oceanogr.*, **21**, 173–180.

Bower, A. S. & Rossby, H. T. (1989). Evidence of cross-frontal exchange processes in the Gulf Stream based on isopycnal RAFOS float data. *J. Phys. Oceanogr.*, **19**, 1177–1190.

Bower, A. S., Rossby, T., & O'Gara, R. (1986). RAFOS float pilot studies in the Gulf Stream. Technical report, University of Rhode Island Tech Report 86-7.

Bowman, K. P. (1993). Barotropic simulation of large-scale mixing in the Antarctic polar vortex. *J. Atmos. Sci.*, **50(17)**, 2901–2914.

Bowman, K. P. (2000). Manifold geometry and mixing in observed atmospheric flows. *preprint*.

Bridges, T. J. & Reich, S. (2001). Computing Lyapunov exponents on a Stiefel manifold. *Physica D*, **156**, 219–238.

Broer, H. W., Huitema, G. B., & Sevryuk, M. B. (1996). *Quasi-Periodic Motions in Families of Dynamical Systems*, volume 1645 of *Lecture Notes in Mathematics*. Springer-Verlag, New York, Heidelberg, Berlin.

Brown, M. G. & Samelson, R. M. (1994). Particle motion in vorticity-conserving, two-dimensional incompressible flows. *Phys. Fluids*, **6(9)**, 2875–2876.

Camassa, R. & Wiggins, S. (1991). Chaotic advection in a Rayleigh-Bénard flow. *Phys. Rev. A*, **43(2)**, 774–797.

Cartwright, J. H. E., Feingold, M., & Piro, O. (1994). Passive scalars and 3-dimensional Liouvillian maps. *Physica D*, **76(1–3)**, 22–33.

Cartwright, J. H. E., Feingold, M., & Piro, O. (1995). Global diffusion in a realistic 3-dimensional time-dependent nonturbulent fluid flow. *Phys. Rev. Lett.*, **75(20)**, 3669–3672.

Cartwright, J. H. E., Feingold, M., & Piro, O. (1996). Chaotic advection in three-dimensional unsteady incompressible laminar flow. *J. Fluid. Mech.*, **316**, 259–284.

Casasayas, J., Fontich, E., & Nunes, A. (1992). Invariant manifolds for a class of parabolic points. *Nonlinearity*, **5**, 1193–1210.

Celletti, A. & Chierchia, L. (1988). Construction of analytic KAM surfaces and effective stability bounds. *Comm. Math. Phys.*, **118**, 119–161.

Cencini, M., Lacorata, G., Vulpiani, A., & Zambianchi, E. (1999). Mixing in a meandering jet: A Markovian approximation. *J. Phys. Oceanogr.*, **29(10)**, 2578–2594.

Channon, S. R. & Lebowitz, J. L. (1980). Numerical experiments in sto-chasticity and homoclinic oscillations. *Ann. New York Acad. Sci.*, **357**, 108.

Chicone, C. (1999). *Ordinary Differential Equations with Applications*. Springer-Verlag, New York.

Clever, R. H. & Busse, F. (1974). Transition to time-dependent convection. *J. Fluid Mech.*, **65**, 625–645.

Coddington, E. A. & Levinson, N. (1955). *Theory of Ordinary Differential Equations*. McGraw-Hill, New York.

Coppel, W. A. (1978). *Dichotomies in Stability Theory*, volume 629 of *Lecture Notes in Mathematics*. Springer-Verlag, New York, Heidelberg, Berlin.

Coulliette, C. & Wiggins, S. (2001). Intergyre transport in a wind-driven, quasigeostrophic double gyre: An application of lobe dynamics. *Nonlinear Processes in Geophysics*, **8**, 69–94.

Cronin, M. & Watts, D. R. (1996). Eddy-mean flow ineraction in the Gulf Stream at 68 degrees W.1.1 Eddy energetics. *J. Phys. Oceanogr.*, **26(10)**, 2107–2131.

de Blasi, F. S. & Schinas, J. (1973). On the stable manifold theorem for discrete time dependent processes in Banach spaces. *Bull. London Math. Soc.*, **5**, 275–282.

de la Llave, R. & Rana, D. (1990). Accurate strategies for small divisor problems. *Bull. Am. Math. Soc.*, **22(1)**, 85–90.

Deese, H. E., Pratt, L. J., & Helfrich, K. R. (2002). A laboratory model of exchange and mixing between western boundary layers and subbasin recirculation gyres. *J. Phys. Oceanogr.*, **32(6)**, 1870–1889.

del Castillo-Negrete, D. & Morrison, P. J. (1993). Chaotic transport by Rossby waves in shear flow. *Phys. Fluids A*, **5(4)**, 948–965.

del Castillo-Negrete, D., Greene, J. M., & Morrison, P. J. (1996). Area-preserving nontwist maps: Periodic orbits and transition to chaos. *Physica D*, **91(1–2)**, 1–23.

del Castillo-Negrete, D., Greene, J. M., & Morrison, P. J. (1997). Renormalization and transition to chaos in area preserving ontwist maps. *Physica D*, **100(3–4)**, 311–329.

Desoer, C. A. (1969). Slowly varying $\dot{x} = a(t)x$. *IEEE Trans. Automat. Control*, **14**, 1091–1100.

Dieci, L. & Eirola, T. (1999). On smooth decompositions of matrices. *SIAM J. Matrix Anal. Appl.*, **20(3)**, 800–819.

Dieci, L. & Vleck, E. S. V. (1999). Computation of orthonormal factors for fundamental solution matrices. *Numer. Math.*, **83**, 599–620.

Dieci, L., Russell, R. D., & Vleck, E. S. V. (1997). On the computation of Lyapunov exponents for continuous dynamical systems. *SIAM J. Numer. Anal.*, **34(1)**, 402–423.

d'Ovidio, F., Fernández, V., Hernández-Garcia, E., & López, C. (2004). Mixing structures in the Mediterranean sea from finite size Lyapunov exponents. *Geophysical Research Letters*, **31**, L17203.

Drazin, P. G. & Reid, W. H. (1981). *Hydrodynamic stability*. Cambridge University Press, Cambridge.

Dritschel, D. (1989). Contour dynamics and contour surgery: Numerical algorithms for extended, high-resolution modelling of vortex dynamics in two-dimensional, inviscid, incompressible flows. *Comp. Phys. Rep.*, **10**, 77–146.

Dritschel, D. G. & Ambaum, M. H. P. (1997). A contour-advective semi-Lagrangian numerical algorithm for simulating fine-scale conservative dynamical fields. *Q. J. R. Meteorol. Soc.*, **123**, 1097–1130.

Duan, J. & Wiggins, S. (1997). Lagrangian transport and chaos in the near wake of the flow around an obstacle: a numerical implementation of lobe dynamics. *Nonlinear Processes in Geophysics*, **4**, 125–136.

Duan, J. Q. & Wiggins, S. (1996). Fluid exchange across a meandering jet with quasi-periodic time variability. *J. Phys. Oceanogr.*, **26**, 1176–1188.

Dutkiewicz, S., Griffa, A., & Olson, D. B. (1993). Particle diffusion in a meandering jet. *J. Geophys. Res.*, **98(C9)**, 16487–16500.

Farrell, B. F. & Ioannou, P. J. (1996). Generalized stability theory. part I: Autonomous operators. *J. Atmos. Sci.*, **53(14)**, 2025–2040.

Fontich, E. (1999). Stable curves asymptotic to a degenerate fixed point. *Nonlin. Anal.*, **35**, 711–733.

Fountain, G. O., Khakhar, D. V., Mezic, I., & Ottino, J. M. (2000). Chaotic mixing in a bounded three-dimensional flow. *J. Fluid Mech.*, **417**, 265–301.

Geist, K., Parlitz, U., & Lauterborn, W. (1990). Comparison of different methods for computing Lyapunov exponents. *Prog. Theor. Phys.*, **83(5)**, 875–893.

Goldhirsch, I., Sulem, P.-L., & Orszag, S. A. (1987). Stability and Lyapunov stability of dynamical systems: A differential approach and a numerical method. *Physica D*, **27**, 311–337.

Goldstein, H. (1980). *Classical Mechanics*. Addison-Wesley, Reading, MA, second edition.

Greene, J. M. & Kim, J.-S. (1987). The calculation of Lyapunov spectra. *Physica D*, **24**, 213–225.

Guckenheimer, J. & Holmes, P. (1983). *Nonlinear Oscillations, Dynamical Systems, and Bifurcations of Vector Fields*. Springer-Verlag, New York.

Hale, J. (1980). *Ordinary Differential Equations*. Robert E. Krieger Publishing Co., Inc, Malabar, Florida.

Haller, G. (2000). Finding finite-time invariant manifolds in two-dimensional velocity fields. *Chaos*, **10(1)**, 99–108.

Haller, G. (2001). Distinguished material surfaces and coherent structure in three-dimensional fluid flows. *Physica D*, **149**, 248–277.

Haller, G. (2002). Lagrangian coherent structures from approximate velocity data. *Phys. Fluids*, **14(6)**, 1851–1861.

Haller, G. & Mezic, I. (1998). Reduction of three-dimensional, volume-preserving flows with symmetry. *Nonlinearity*, **11(2)**, 319–339.

Haller, G. & Poje, A. (1998). Finite time transport in aperiodic flows. *Physica D*, **119**, 352–380.

Haller, G. & Yuan, G. (2000). Lagrangian coherent structures and mixing in two-dimensional turbulence. *Physica D*, **147**, 352–370.

Henry, D. (1981). *Geometric theory of semilinear parabolic equations, Lecture Notes in Mathematics, Vol. 840*. Springer-Verlag: New York, Heidelberg, Berlin.

Herman, M. R. (1988). Existence et non existence de tores invariants par des difféomorphismes symplectiques. *preprint*.

Hirsch, M. W. & Smale, S. (1974). *Differential Equations, Dynamical Systems, and Linear Algebra*. Academic Press, New York.

Holton, J. (1992). *An Introduction to Dynamic Meteorology, 3rd ed.* Academic Press, San Diego.

Howard, J. E. & Humphreys, J. (1995). Nonmonotonic twist maps. *Physica D*, **80**, 256–276.

Ide, K., Small, D., & Wiggins, S. (2002). Distinguished hyperbolic trajectories in time dependent fluid flows: analytical and computational approach for velocity fields defined as data sets. *Nonlinear Processes in Geophysics*, **9**, 237–263.

Irwin, M. C. (1973). Hyperbolic time dependent processes. *Bull. London Math. Soc.*, **5**, 209–217.

Janaki, T. M., Rangarajan, G., Habib, S., & Ryne, R. D. (1999). Computation of the Lyapunov spectrum for continuous-time dynamical systems and discrete maps. *Phys. Rev. E.*, **60(6)**, 6614–6626.

Jorba, A. & Simo, C. (1996). On quasiperiodic perturbations of elliptic equilibrium points. *SIAM J. Math. Anal.*, **27(6)**, 1704–1737.

Joseph, B. & Legras, B. (2002). Relation between kinematic boundaries, stirring, and barriers for the Antarctic polar vortex. *J. Atmos. Sci.*, **59(7)**, 1198–1212.

Ju, N., Small, D., & Wiggins, S. (2003). Existence and computation of hyperbolic trajectories of aperiodically time-dependent vector fields and their approximations. *Int. J. Bif. Chaos*, **13**, 1449–1457.

Kaper, T. J. (1992). *On the structure of separatrix swept regions of slowly modulated Hamiltonian systems. On the quantification of mixing in chaotic Stokes flows: the eccentric journal bearing*. Ph.D. thesis, California Institute of Technology.

Kaper, T. J. & Wiggins, S. (1991). Lobe area in adiabatic Hamiltonian systems. *Physica D*, **51 (1-3)**, 205–212.

Kaper, T. J. & Wiggins, S. (1992). On the structure of separatrix swept regions in singularly perturbed Hamiltonian systems. *Differential and Integral Equations*, **5(6)**, 1363–1381.

Katok, A. & Hasselblatt, B. (1995). *Introduction to the Modern Theory of Dynamical Systems.* Cambridge University Press, Cambridge.

Knobloch, E. & Weiss, J. B. (1987). Chaotic advection by modulated traveling waves. *Phys. Rev. A*, **36**, 1522–1524.

Koh, T.-Y. & Plumb, R. A. (2000). Lobe dynamics applied to barotropic Rossby wave breaking. *Phys. Fluids*, **12(6)**, 1518–1528.

Kolmogorov, A. N. (1954). On conservation of conditionally periodic motions under small perturbations of the Hamiltonian. *Dokl. Akad. Nauk. USSR*, **98(4)**, 527–530.

Kovacic, G. (1991). Lobe area via action formalism in a class of Hamiltonian systems. *Physica D*, **51 (1–3)**, 226–233.

Kozlov, V. V. (1996). *Symmetries, Topology and Resonances in Hamiltonian Mechanics.* Springer-Verlag, New York.

Krauskopf, B., Osinga, H. M., Doedel, E. J., Henderson, M. E., Guckenheimer, J., Vladimirsky, A., Dellnitz, M., & Junge, O. (2005). A survey of methods for computing (un)stable manifolds of vector fields. *Int. J. Bif. Chaos*, **15**(3), 763–791.

Landau, L. D. & Lifschitz, E. M. (1976). *Mechanics.* Pergamon, Oxford.

Lapeyre, G. (2002). Characterization of finite-time Lyapunov exponents and vectors in two-dimensional turbulence. *Chaos*, **12(3)**, 688–698.

Lee, T. (1994). *Variability of the Gulf Stream path observed from satellite infrared images.* Ph.D. thesis, University of Rhode Island.

Lee, T. & Cornillon, P. (1995). Temporal variation of meandering intensity and domain-wide lateral oscillations of the Gulf Stream. *J. Geophys. Res.*, **100**, 13603–13613.

Legras, B. & Vautard, R. (1996). A guide to Liapunov vectors. In T. Palmer, editor, *Proceedings of the 1995 ECMWF Seminar on Predictability*, pages 143–156.

Lozier, M. S. & Bercovici, D. (1992). Particle exchange in an unstable jet. *J. Phys. Oceanogr.*, **22**, 1506–1516.

Lozier, M. S., Pratt, L. J., Rogerson, A. M., & Miller, P. D. (1997). Exchange geometry revealed by float trajectories in the Gulf Stream. *J. Phys. Oceanogr.*, **27**, 2327–2341.

MacKay, R. S. (1990). A criterion for non-existence of invariant tori for Hamiltonian systems. *Physica D*, **36(1–2)**, 64–82.

MacKay, R. S. & Percival, I. C. (1985). Converse KAM: Theory and practice. *Comm. Math. Phys.*, **98**, 469–512.

MacKay, R. S., Meiss, J. D., & Percival, I. C. (1984). Transport in Hamiltonian systems. *Physica D*, **13**, 55–81.

MacKay, R. S., Meiss, J. D., & Stark, J. (1989). Converse KAM theory for symplectic twist maps. *Nonlinearity*, **2**, 555–570.

Malhotra, N. & Wiggins, S. (1998). Geometric structures, lobe dynamics, and Lagrangian transport in flows with aperiodic time-dependence, with applications to Rossby wave flow. *J. Nonlinear Science*, **8**, 401–456.

Mancho, A., Small, D., Wiggins, S., & Ide, K. (2003). Computation of stable and unstable manifolds of hyperbolic trajectories in two-dimensional, aperiodically time-dependent vector fields. *Physica D*, **182**, 188–222.

Mancho, A., Small, D., & Wiggins, S. (2004). Computation of hyperbolic and their stable and unstable manifolds for oceanographic flows represented as data sets. *Nonlinear Processes in Geophysics*, **11(1)**, 17–33.

Mancho, A., Small, D., & Wiggins, S. (2006). A tutorial on dynamical systems concepts applied to Lagrangian transport in oceanic flows defined as finite time data sets: Theoretical and computational issues. *Submitted to Physics Reports*.

Massera, J. L. & Schäffer, J. (1966). *Linear differential equations and function spaces*. Academic Press: New York.

Mather, J. (1982). Existence of quasi-periodic orbits for twist maps of the annulus. *Topology*, **21(4)**, 457–467.

Mather, J. (1984). Non-existence of invariant circles. *Ergodic Theory and Dynamical Systems*, **4**, 301–311.

McGehee, R. (1973). A stable manifold theorem for degenerate fixed points with applications to celestial mechanics. *J. Diff. Eq.*, **14**, 70–88.

Melnikov, V. K. (1963). On the stability of the center for time periodic perturbations. *Trans. Moscow Math. Soc.*, **12**, 1–57.

Meyers, S. D. (1994). Cross-frontal mixing in a meandering jet. *J. Phys. Oceanogr.*, **24**, 1641–1646.

Mezic, I. (2001a). Break-up of invariant surfaces in action-angle-angle maps and flows. *Physica D*, **154(1-2)**, 51–67.

Mezic, I. (2001b). Chaotic advection in bounded Navier-Stokes flow. *J. Fluid Mech.*, **431**, 347–370.

Mezic, I. & Wiggins, S. (1994). On the integrability and perturbation of three-dimensional fluid flows with symmetry. *J. Nonlinear Sci.*, **4**, 157–194.

Mezic, I., Leonard, A., & Wiggins, S. (1998). Regular and chaotic particle motion near a helical vortex filament. *Physica D*, **111(1–4)**, 179–201.

Miller, P. D., Jones, C. K. R. T., Rogerson, A. M., & Pratt, L. J. (1997). Quantifying transport in numerically generated velocity fields. *Physica D*, **110**, 105–122.

Miller, P. D., Pratt, L. J., Helfrich, K. R., & Jones, C. K. R. T. (2002). Chaotic transport of mass and potential vorticity for an island recirculation. *J. Phys. Oceanogr.*, **32(1)**, 80–102.

Month, M. & Herrera, J., editors (1979). *Variational principles for invariant tori and cantori*, volume 57 of *Nonlinear Dynamics and the Beam-Beam Interaction*. Am. Inst. of Phys. Conf. Proc.

Moser, J. (1962). On invariant curves of an area preserving mappings of an annulus. *Nachr. Akad. Wiss. Gött., II. Math.-Phys. Kl.*, pages 1–20.

Munk, W. (1950). On the wind-driven ocean circulation. *J. Meteorology*, **7**, 79–93.

Ngan, K. & Shepherd, T. G. (1997). Chaotic mixing and transport in Rossby wave critical layers. *J. Fluid. Mech.*, **334**, 315–351.

Oseledec, V. (1968). A multiplicative ergodic theorem: Lyapunov characteristic numbers for dynamical systems. *Trans. Moscow Math. Soc.*, **19**, 197–231.

Ottino, J. (1989). *The Kinematics of Mixing: Stretching, Chaos, and Transport.* Cambridge University Press, Cambridge.

Parker, T. S. & Chua, L. O. (1989). *Practical Numerical Algorithms for Chaotic Systems.* Springer-Verlag, Berlin.

Partovi, H. (1999). Reduced tangent dynamics and Lyapunov spectrum for Hamiltonian systems. *Phys. Rev. Lett.*, **82(17)**, 3424–3427.

Pedlosky, J. (1987). *Geophysical Fluid Dynamics, 2nd ed.* Springer-Verlag, New York.

Pedlosky, J. (1996). *Ocean Circulation Theory.* Springer-Verlag, New York.

Percival, I. & Richards, D. (1982). *Introduction to Dynamics.* Cambridge University Press, Cambridge.

Petrisor, E. (2001). Nontwist area preserving maps with reversing symmetry groups. *Int. J. Bif. Chaos*, **11**, 497–512.

Petrisor, E. (2002). Reconnection scenarios and the threshold of reconnection in the dynamics of non-twist maps. *Chaos, Solitions, & Fractals*, **14**, 117–127.

Peuteman, J., Aeyels, D., & Sepulchre, R. (2000). Boundedness properties for time-varying nonlinear systems. *SIAM J. Control. Optim.*, **39(5)**, 1408–1422.

Pierrehumbert, R. T. (1991a). Chaotic mixing of tracer and vorticity by modulated traveling Rossby waves. *Geophys. Astrophys. Fluid Dynamics*, **58**, 285–319.

Pierrehumbert, R. T. (1991b). Large-scale horizontal mixing in planetary atmospheres. *Phys. Fluids A*, **3(5)**, 1250–1260.

Poincaré, H. (1892). *Les méthodes nouvelles de la mécanique celeste.* Gauthier-Villars, Paris.

Poje, A. C. & Haller, G. (1999). Geometry of cross-stream mixing in a double-gyre ocean model. *J. Phys. Oceanogr.*, **29(8)**, 1649–1665.

Poje, A. C., Toner, M., Kirwan, A. D., & Jones, C. K. R. T. (2002). Drifter launch strategies based on Lagrangian templates. *J. Phys. Oceanogr.*, **32(6)**, 1855–1869.

Pratt, L. J., Lozier, M. S., & Beliakova, N. (1995). Parcel trajectories in quasigeostrophic jets: Neutral modes. *J. Phys. Oceanogr.*, **25**, 1451–1466.

Ramasubramanian, K. & Sriram, M. N. (2000). A comparative study of computation of Lyapunov spectra with different algorithms. *Physica D*, **139**, 72–86.

Rangarajan, G., Habib, S., & Ryne, R. D. (1998). Lyapunov exponents without rescaling and reorthogonalization. *Phys. Rev. Lett.*, **80(17)**, 3747–3750.

Regier, L. & Stommel, H. (1979). Float trajectories in simple kinematic flows. *Proc. Nat. Acad. Sci. USA*, **76**, 1760–1764.

Rogerson, A. M., Miller, P. D., Pratt, L. J., & Jones, C. K. R. T. (1999). Lagrangian motion and fluid exchange in a barotropic meandering jet. *J. Phys. Oceanogr.*, **29**, 2635–2655.

Rom-Kedar, V. & Wiggins, S. (1990). Transport in two-dimensional maps. *Arch. Rat. Mech. Anal.*, **109**, 239–298.

Rom-Kedar, V., Leonard, A., & Wiggins, S. (1990). An analytical study of transport, mixing, and chaos in an unsteady vortical flow. *J. Fluid Mech.*, **214**, 347–394.

Salmon, R. (1998). *Lectures on Geophysical Fluid Dynamics*. Oxford University Press, Oxford.

Samelson, R. M. (1992). Fluid exchange across a meandering jet. *J. Phys. Oceanogr.*, **22**, 431–440.

Shariff, K., Pulliam, T., & Ottino, J. (1992). A dynamical systems analysis of kinematics in the time-periodic wake of a circular cylinder. In C. Anderson and C. Greengard, editors, *Vortex dynamics and vortex methods, Proc. AMS-SIAM Conf.*, Lectures in Applied Mathematics, 28, pages 613–646. American Mathematical Society: Providence.

Solomon, T. H. & Gollub, J. P. (1988). Chaotic particle transport in time-dependent Rayleigh-Bénard convection. *Phys. Rev. A*, **38**, 6280–6286.

Sommeria, J., Meyers, S. D., & Swinney, H. L. (1989). Laboratory model of a planetary eastward jet. *Nature*, **337**, 58–61.

Stark, J. (1988). An exhaustive criterion for the non-existence of invariant circles for area-preserving twist maps. *Comm. Math. Phys.*, **117**, 177–189.

Stewartson, K. (1981). Marginally stable inviscid flows with critical layers. *IMA J. App. Math.*, **27(2)**, 133–175.

Stommel, H. (1965). *The Gulf Stream*. University of California Press, Berkeley.

Voyatzis, G. & Ichtiaroglou, S. (1999). Degenerate bifurcations of resonant tori in Hamiltonian systems. *Int. J. Bif. Chaos*, **9(5)**, 849–863.

Warn, T. & Gauthier, P. (1989). Potential vorticity mixing by marginally unstable baroclinic disturbances. *Tellus*, **41A**, 115–131.

Warn, T. & Warn, H. (1978). Evolution of a non-linear critical level. *Stud. App. Math.*, **59(1)**, 37–71.

Weiss, J. B. & Knobloch, E. (1989). Mass transport by modulated traveling waves. *Phys. Rev. A*, **40**, 2579–2589.

Wiggins, A. (2005). The dynamical systems approach to Lagrangian transport in oceanic flows. *Annu. Rev. Fluid Mech.*, **37**, 295–328.

Wiggins, S. (1988). *Global Bifurcations and Chaos: Analytical Methods*. Springer-Verlag, New York.

Wiggins, S. (1992). *Chaotic Transport in Dynamical Systems*. Springer-Verlag, New York.

Wiggins, S. (2003). *Introduction to Applied Nonlinear Dynamical Systems and Chaos, second edition.* Springer-Verlag, New York.

Yamaguchi, Y. Y. & Iwai, T. (2001). Geometric approach to Lyapunov analysis in Hamiltonian dynamics. *Phys. Rev. E.*, **64**, 066206.

Yannacopoulos, A. N., Mezic, I., Rowlands, G., & King, G. P. (1998). Eulerian diagnostics for Lagrangian chaos in three-dimensional Navier-Stokes flows. *Phys. Rev. E*, **57(1)**, 482–490.

Yi, Y. (1993). A generalized integral manifold theorem. *J. Diff. Eq.*, **102**, 153–187.

Yuan, G.-C., Pratt, L. J., & Jones, C. K. R. T. (2001). Barrier destruction and Lagrangian predictabilty at depth in a meandering jet. *Dyn. Atmos. Oceans*, **35**, 41–61.

Yuan, G.-C., Pratt, L. J., & Jones, C. K. R. T. (2004). Cross-jet Lagrangian transport and mixing in a $2\frac{1}{2}$ layer model. *J. Phys. Oceanogr.*, **34**(9), 1991–2005.

Yuster, T. & Hackborn, W. W. (1997). On invariant manifolds attached to oscillating boundaries in Stokes flows. *Chaos*, **7**, 769–776.

Index

Interdisciplinary Applied Mathematics